EVOLUTION

CHARLOTTE J. AVERS
Douglass College, Rutgers University

Harper & Row, Publishers

New York Evanston San Francisco London

Sponsoring Editor: *Joe Ingram*
Project Editor: *Robert E. Ginsberg*
Designer: *Frances Torbert Tilley*
Production Supervisor: *Robert A. Pirrung*

Library of Congress Cataloging in Publication Data
Avers, Charlotte J
  Evolution
  1. Evolution. I. Title. [DNLM: 1. Evolution.
QH366.2 A953e 1974]
QH366.2.A93   575   73–9367
ISBN 0–06–040398–5

CONTENTS

v

PART II PRINCIPLES AND PHENOMENA

PART III TRENDS IN EVOLUTION

PREFACE

Evolution is an exciting field of study. It permits us to form a broad view of the universe and of our place in the scheme of things. Such a perspective derives from many kinds of information that can be integrated into a satisfying intellectual framework. Like many other areas of scholarship, the study of evolution draws on methods and information from different approaches and bodies of knowledge. There is input from astronomy, geology, chemistry, physics, mathematics, and anthropology, along with such biological subjects as genetics, physiology, anatomy, biochemistry, and behavior. All together these help us blend the diverse components of evolutionary history to fashion a comprehensive theory of evolution of life on Earth.

The experimental approach underlies the uniqueness of science among the scholarly disciplines. The scientist may be able to recreate situations and systems, under controlled experimental conditions, and study events and phenomena of past and present times and places. For example, we can simulate some conditions of the primeval Earth of billions of years ago and determine the kinds of chemicals produced in reactions significant to the origin of life. By combining data derived from experimentation with geological and fossil evidence and with information from modern life forms, we may come closer to understanding the origin of life than would be possible without insights from experimental analysis. Such experiments do not necessarily prove that similar events happened long ago, but they at least indicate the plausibility of the working hypothesis and encourage us to design further experiments in searching for definitive answers to questions about evolution on Earth.

We can also examine the fossil record and, interpreting the information against a background of sufficient knowledge, construct hypotheses about the progression of life forms and of the biological processes that influenced these sequences. As we understand the system more fully we can make predictions and continue to fill in the outlines and details of evolutionary history. We thus interpret the fossil record as showing that life was absent from the Earth in the beginning, that simple forms appeared first, and that complex life evolved from these simpler ancestors. We predict that additional fossil finds will fit the proposed pattern of evolution. And as newer discoveries do fit the predicted pattern, the evolutionary theory is strengthened. It is not necessary or feasible to recreate the Earth and its life entirely in order to verify current ideas about evolutionary progressions. Both the experimental and historical approaches can provide information about our evolutionary past. The combination of experimental and descriptive analysis has proven fruitful in guiding scholarly inquiry into the history of life on our planet.

This book has been arranged in four parts. The first chapters cover the themes and story of relationships between life and nonlife in the universe and on Earth. From an examination of the composition and origin of the universe we proceed to the unity of matter and the prerequisites for life. After we investigate how life originated, we look at the development of the genetic and metabolic systems that underlie the unity and diversity of all living things. In the second part we discuss principles of evolutionary processes and phenomena of evolutionary change. The recurring and universal themes of mutation and natural selection are discussed more thoroughly, and the concept of evolution at the population level is developed. The second section concludes with a chapter on the evolution of genetic systems, including sexual reproduction and chromosome patterns. Certain trends in evolution are discussed in the third group of chapters, including patterns of speciation, extinction, and adaptation. The geological and fossil records are outlined and some attention is directed toward recent evidence showing that continental migrations occur and influence species patterns. A chapter on particular features of vertebrate evolution also serves to introduce the reader to the primates, our ancestors and closest relatives. In the last set of chapters we are specifically concerned with human evolution, including our immediate ancestry, patterns of social organization, and promises for the future. In this section the emphasis is placed on relationships of the human family to other primate groups. The similarities and differences between us and our primate relatives provide a format for interpreting our uniqueness as a product of a particular evolutionary sequence.

I have been greatly aided in writing this book by assimilating the ideas and responses of students in the Evolution course I have taught for the past fifteen years. As I have stressed in these classes, so I have tried to stress in this book the lines of thinking and reasoning that are involved, how we obtain and interpret evidence of evolutionary processes and phenomena, and the logical way in which inquiry is conducted. The challenge and excitement of intellectual inquiry and problem-solving have maintained my interest in the study of evolution. I hope that some of this challenge and stimulation will be communicated to the reader.

CHARLOTTE J. AVERS

November, 1973
New Brunswick, New Jersey

PART I

LIFE
AND
PRELIFE

CHAPTER 1
INTRODUCTORY REMARKS

Evolution implies change with time, both for living and for nonliving systems. We are aware of the processes of change in life forms during the billions of years of the Earth's history but perhaps do not quite realize the phenomena of change that also characterize the inanimate world. Changes brought about by the gradual erosion of rocks, mountains, and hillsides are aspects of evolution in that they occur inexorably over periods of time. The stars in the vastness of space are evolving as we look at them. The star shines as a consequence of energy radiating from the mass of the matter of which the star is composed. These radiations provide information about the kinds of atomic reactions taking place. As the temperature of the star changes with time, the kinds of reactions change and the composition of the star changes; evolution in action is observable on any clear night. These sequences of change can be interpreted in terms of the "aging" of the star, that is, of a series of events that differ during the existence of the star from its beginning to some predictable endpoint when radiations stop all together.

There is a significant difference between evolution of life forms and evolution of inanimate matter. For all nonliving systems about which we know anything, the sequences of change are predictable. Such systems undergo alterations in specific and determined ways, as in the example of the star. Once its fuel supply is exhausted, the star dies. For living systems, however, change is unpredictable. We certainly may predict that changes will occur, but we cannot predict what those changes will be. We look back at the fossil record and can determine that reptiles gave rise to entirely new kinds of life that we now know as the birds and the mammals, but we are unable to say what new forms will arise in the future or from what ancestral lines these new forms will diverge.

Another way in which living and nonliving systems differ is in the quality of repeatability. Evolutionary sequences occur repeatedly in nonlife systems, and there is little variation from one set of events to the next. Stars are born, evolve, and die. With little difference between different stars except for the initial components out of which the mass of matter forms, we can expect similar sequences of change in almost any star of comparable age and composition. Indeed, knowing how similar the stars are as a particular nonlife system, it is possible to construct hypotheses that relate sets of events to any star, whether in existence now or in the past or in the future. Living systems do not repeat evolutionary sequences, except for superficially similar forms such as the flying insects, birds, and bats, or the swimming fish and whale. The evolution of flight occurred by quite different routes in insects and in birds; we cannot consider the appearance of birds as a repeat of the evolutionary sequence of events that happened in insect populations. New forms appear regularly during the evolution of life on the planet, but new aspects of nonliving matter rarely occur on the Earth. Interestingly, humans have been able to create entirely new kinds of inanimate matter in recent times, as in the case of synthetic plastics. Most other kinds of nonliving materials have been present on the Earth for billions of years in essentially the same composition and format during all this time.

Another way in which life and nonlife differ is in the relationship between the parent materials and the final products. Living systems produce new forms very much like themselves at first, and the changes accumulate until another species or family of organisms has evolved. The lineage can be traced to show very clearly the intimate relationship between ancestral and derived forms, both sharing more features in common than differences. In nonliving systems the parent materials can be quite unrelated to the final products. Dust and gases of interstellar space may condense to form a star that is very different in chemical composition, temperature, reaction spectrum, and conformation relative to the materials from which it formed. Of all the qualities of the evolution of life, the most dramatic and significant feature is that of unpredictability. The basis of understanding evolution includes this peculiar aspect of change of living forms and its implications when viewing modern species.

There are many ways to study evolution, all of which are important in the total picture that emerges. The older approaches, which still are valid and highly desirable, include the methods of historical geology and comparative analysis of existing species. Fossils embedded in the rocks of the planet provide the principal tangible evidence of former life. The fossil record indicates a progression of species during several billion years of time, with obvious similarities due to relationship and common ancestry that can be traced in many cases. Verification of descent with modification, as seen in the fossil record, is obtained from comparative morphology, comparative anatomy, comparative biochemistry, comparative genetics, and other lines of investigation which permit us to determine lineage and the degree of relationship among present-day life forms. In most cases the evidence clearly supports the deductions from the fossil record and the variety of experimental and observational data to show that the processions of life forms make up series of lineages of descent. The successions of modifications lead to infinite variations on original patterns, but the patterns can be discerned and the relationship of all life can be traced back to very early beginnings. These lines of evidence provide the overwhelming support for the theory of evolution as descent with modification, change with time from ancestral forms to those that exist on the Earth today.

The fossil record and comparative studies alone, however, cannot provide all the information necessary to determine how evolution has occurred or why particular pathways of change have been utilized. Additional information is needed to provide the full perspective of biological evolution in its most dynamic aspects. The evidence from the fossil record and from comparative analysis indicate *what* happened and *when* it happened but relatively little of the *how* and *why* of evolution.

Some Historical Background

Our concepts of the physical universe have changed frequently during the millenia of recorded history, but modern views of the universe were developed principally during the twentieth century. Whereas people in the Middle Ages considered humans to be the pivotal component of the universe and therefore placed the Earth at the very center, we have a humbler view of ourselves today. We know that the Earth is but one of nine planets orbiting an average star in one of many solar systems. Our sun is considered to be an average star in age and size, one of 100 billion stars in our galaxy, the Milky Way. The solar system we inhabit is near the edge of the galaxy, not near the center of this vastness at all. The Milky Way is only one of the billion galaxies in the universe. Since our solar system may be just one of the million estimated to be present in our own galaxy, and since similar solar systems are believed to occur in all galaxies, there may be a million billion solar systems and their constituent planets in the universe.

Such staggering numbers of entities in the universe are made even more amazing upon consideration of the distances that separate them. Distances are discussed in terms of light-years. One light-year is equivalent to the distance traveled by light in one year, moving at the rate of 186,000 miles per second. In more familiar terms, this distance equals about 6 trillion miles. Using this form of measurement, we find that the diameter of the Milky Way galaxy is about 100,000 light-years or about 600 million billion miles across. It takes 30,000 years for light from our sun to reach the center of the galaxy, traveling at 186,000 miles per second. In other words, the sun is 30,000 light-years distant from the center of the Milky Way. Our nearest star neighbor is about 4 light-years distant, or about 24 trillion miles away.

Reflecting telescopes, such as those on Mount Palomar, permit us to see about 1 or 2 billion light-years into space, but with radio telescopes some believe that we can hear signals from sources as far away as 8 or even 10 billion light-years distant from us in the universe. Interstellar space once was considered to be a vacuum, but now we know that matter exists in areas between the stars principally as a thin scattering of elements everywhere, but occasionally in the form of clouds of dust and gases. Interstellar material can be studied only with appropriate and sensitive instruments.

Our understanding and knowledge of the nature and structure of matter has expanded from the seventeenth century to the present day. Boyle proposed the Corpuscular Theory of Matter in the seventeenth century, and the concepts of atoms and molecules were developed later by Dalton and Avogadro. The first organic compound was synthesized by Wöhler in the mid-nineteenth century, indicating the continuity of matter in living and nonliving systems. The synthesis of the organic chemical urea provided evidence of the bridge that unites the animate and inanimate materials of our world and the universe. Numerous significant advances from the field of physics increased our comprehension of the atom and its component parts, of the interconvertibility of energy and matter as presented in the familiar Einstein equation of $E = mc^2$, and of the nature of radiations, subatomic particles, and many other phenomena of the natural world. Similarly, developments in biology were important to evolutionary theory. The development of increasingly sophisticated instrumentation, beginning with the invention of the microscope in the seventeenth century, was essential to advancement. The demonstration by Pasteur that life came only from preexisting life marked a milestone, as did the contributions made by Lamarck and Darwin and Wallace to evolutionary theory. Modern studies have widened our perspective by incorporating information about genetics and the concept of the population into a framework we term neo-Darwinian evolutionary theory.

Dating Methods

The development of evolutionary theory depended in large part upon our understanding of time and upon the means for measuring time. The calculations made by Archbishop Ussher, a nineteenth-century fundamentalist cleric, which showed that the Earth was formed on October 23, 4043 B.C., are quite different from more reliable calculations that place the age of the Earth at 4 billion 6 hundred million (4,600,000,000) years.

Before the invention of particle accelerators such as cyclotrons and bevatrons, we knew of 92 elements, from the lightest, hydrogen, to uranium as the heaviest in the sequence (Fig. 1.1). A number of elements heavier than uranium have been produced in the laboratory, many of which have a fleeting existence. Elements are composed of subatomic particles. Many kinds of subatomic particles have been described, including stable and unstable types, matter and antimatter, those that interact with other particles and those that seldom collide with or influence other particles, in a bewildering variety of sizes. For our purposes, we can concentrate on three principal subatomic particles, namely, protons, neutrons, and electrons. Protons have a mass of 1 and a positive electrical charge; neutrons also have a mass of 1 but lack an electrical charge. These are the two main constituents of the nucleus of the atom (Fig. 1.2) and are the basis for the mass and charge characteristics of the elements. The charge of the element is a function of the protons in the atomic nucleus and is indicated by a subscript that designates the *atomic number* of the element in the periodic table. The mass of an element is derived from the combined numbers of protons and neutrons of the nucleus and is indicated by a superscript that denotes the *atomic weight* of the element. Thus, uranium is designated $^{238}_{92}U$, showing that there are in the nucleus 92 protons plus 146 neutrons.

The third important kind of subatomic particle is the electron. These are very small negatively charged components, which swarm around the nucleus of the atom. There is a specified number of electrons for each element. A similar particle carrying a positive electrical charge is known as a positron.

The isotopes of an element all have the same charge (number of protons) but differ in mass because there are different numbers of neutrons in each isotope of the same element. The isotopes of hydrogen include the commoner ^1H, ^2H (known as deuterium and a component of the "heavy" water used in thermonuclear devices), plus ^3H or tritium, which is a radioactive isotope of considerable importance in biological tracer studies. A radioactive element is relatively unstable and emits radiation, which can be detected by proper procedures and instrumentation.

The conversion of one isotope to another, or of one element to another, can occur by a process known as neutron capture as well as by other means. The sequence of change from hydrogen to helium, which is of great importance in the atomic reactions in stars, may occur by neutron capture according to the following scheme:

$$^1_1H \xrightarrow{+n} {}^2_1H \xrightarrow{+n} {}^3_1H \xrightarrow{+n} {}^4_1H \xrightarrow{-e^-} {}^4_2He$$

hydrogen   deuterium   tritium   [unstable]   helium

The addition of a third neutron leads to an unstable atom, which, after the loss of a negatively charged electron, converts one of the neutrons to a proton. The charge is increased because there now are two protons in the nucleus along with two neutrons, which remain unchanged. The new element is helium, with a mass of 4 and an atomic number of 2. Some of these reactions and concepts are important in those dating methods that utilize starting materials and final products of atomic reactions to arrive at an estimate of the age of some form of matter.

Radioactive decay series involve reactions presumed to occur at rates that are independent of temperature, pressure, and particular combinations

LIGHT METALS

I A	II A
1 **H** Hydrogen 1.0080	
3 **Li** Lithium 6.939	4 **Be** Beryllium 9.012
11 **Na** Sodium 22.990	12 **Mg** Magnesium 24.31

KEY

26	ATOMIC NUMBER
Fe	ELEMENT SYMBOL
Iron	ELEMENT NAME
55.85	ATOMIC WEIGHT

HEAVY METALS

		III B	IV B	V B	VI B	VII B			VIII B	
19 **K** Potassium 39.102	20 **Ca** Calcium 40.08	21 **Sc** Scandium 44.96	22 **Ti** Titanium 47.90	23 **V** Vanadium 50.94	24 **Cr** Chromium 52.00	25 **Mn** Manganese 54.94	26 **Fe** Iron 55.85	27 **Co** Cobalt 58.93		
37 **Rb** Rubidium 85.47	38 **Sr** Strontium 87.62	39 **Y** Ytterbium 88.91	40 **Zr** Zirconium 91.22	41 **Nb** Niobium 92.91	42 **Mo** Molyb-denum 95.94	43 **Tc** Tech-netium (99)	44 **Ru** Ruthenium 101.1	45 **Rh** Rhodium 102.90		
55 **Cs** Cesium 132.91	56 **Ba** Barium 137.34	57 to 71	72 **Hf** Hafnuim 178.49	73 **Ta** Tantalum 180.95	74 **W** Tungsten 183.85	75 **Re** Rhenium 186.2	76 **Os** Osmium 190.2	77 **Ir** Iridium 192.2		
87 **Fr** Francium (223)	88 **Ra** Radium 226.05	89 to 103								

	57 **La** Lanthanum 138.91	58 **Ce** Cerium 140.12	59 **Pr** Prase-odymium 140.91	60 **Nd** Neo-dymium 144.24	61 **Pm** Prome-thium (147)	62 **Sm** Samarium 150.35	63 **Eu** Europium 151.96
LANTHANIDE SERIES							
ACTINIDE SERIES	89 **Ac** Actinum (227)	90 **Th** Thorium 232.04	91 **Pa** Protacti-nium (231)	92 **U** Uranium 238.03	93 **Np** Neptunium (237)	94 **Pu** Plutonium (242)	95 **Am** Americium (243)

Figure I.I   Table of the elements based on periodic classification.

of atoms as molecules. Because of these attributes, there is a constant rate of the loss of initial material and the accumulation of endproducts and, therefore, a constant proportion of the two sets of elements. Because there is a constant proportion of starting materials and endproducts with respect to time, such decay series may be used to determine the time elapsed or the age of a sample containing the elements of the decay series. Although the

(Based on C^{12} = 12.0000)
1961 Atomic Weights

								VIII A
				NONMETALS				2 He Helium 4.003
			IIIA	IV A	V A	VI A	VIIA	
			5 B Boron 10.81	6 C Carbon 12.011	7 N Nitrogen 14.007	8 O Oxygen 15.994	9 F Fluorine 18.998	10 Ne Neon 20.183
	I B	II B	13 Al Aluminum 26.98	14 Si Silicon 28.09	15 P Phos- phorus 30.974	16 S Sulfur 32.064	17 Cl Chlorine 35.453	18 Ar Argon 39.948
28 Ni Nickel 58.71	29 Cu Copper 63.54	30 Zn Zinc 65.37	31 Ga Gallium 69.72	32 Ge Germa- ium 72.59	33 As Arsenic 74.92	34 Se Selenium 78.96	35 Br Bromine 79.909	36 Kr Krypton 83.80
46 Pd Palladium 106.4	47 Ag Silver 107.870	48 Cd Cadmium 112.40	49 In Indium 114.82	50 Sn Tin 118.69	51 Sb Antimony 121.75	52 Te Tellurium 127.60	53 I Iodine 126.90	54 Xe Xenon 131.30
78 Pt Platinum 195.09	79 Au Gold 197.0	80 Hg Mercury 200.59	81 Tl Thallium 204.37	82 Pb Lead 207.19	83 Bi Bismuth 208.98	84 Po Polonium (210)	85 At Astatine (210)	86 Rn Radon (222)

64 Gd Gado- linium 157.25	65 Tb Terbium 158.92	66 Dy Dyspro- sium 162.50	67 Ho Holmium 164.93	68 Er Erbium 167.26	69 Tm Thulium 168.93	70 Yb Ytterbium 173.04	71 Lu Lutetium 174.97
96 Cm Curium (247)	97 Bk Berkelium (249)	98 Cf Calif- ornium (251)	99 Es Einstein- ium (254)	100 Fm Fermium (253)	101 Md Mendel- evium (256)	102 No Nobelium (254)	103 Lw Law- rencium (257)

decay rates of elements vary from microseconds to billions of years, one can determine the *half-life* of an element as the time it takes for half the amount of starting material to be changed to endproducts. Because it is important to know only the proportion of the starting and endproduct substances, the absolute amounts of these materials are not essential for the calculations.

DECAY SERIES USED TO DATE INORGANIC MATERIALS

Three examples of decay series of this sort are uranium–lead, potassium–argon, and rubidium–strontium. In the uranium–lead decay series, it is important to know that there are four isotopes of the element lead (Pb), all of which occur in specific proportions to each other in normal lead deposits. One of these isotopes, ^{204}Pb, is not a product of radioactive decay and is therefore assumed to have been present and unchanged since the formation of the Earth's crust. The remaining three isotopes are derived as follows: ^{206}Pb from ^{238}U, ^{207}Pb from ^{235}U, and ^{208}Pb from ^{232}Th. Upon fission of the atomic nucleus of ^{238}U, there is a loss of electrons and the formation of lead and helium according to the following reaction:

$$^{238}_{92}U \xrightarrow{\text{fission}} {}^{206}_{82}Pb + 8\,{}^{4}_{2}He + 6\,e^-$$

The half-lives of the uranium isotopes and of thorium are as follows:

^{235}U, 1.03×10^9 years*
^{238}U, 6.51×10^9 years
^{232}Th, 20.10×10^9 years

Knowing the half-lives of these elements, one then determines the present ratio of lead isotopes in a sample, plus the amount of ^{204}Pb which serves as a constant value. A simple algebraic equation with three unknowns is set up and then solved to learn the age of the rock sample in which the elements occur.

In the potassium–argon series, ^{40}K decays to ^{40}A + ^{40}Ca, with a half-life of 1.3×10^9 years. For rubidium–strontium, ^{87}Rb decays to ^{87}Sr with a half-life of 9×10^8 years.

Datings of rocks from various parts of the world have revealed a considerable range, some rocks being almost 4 billion years of age. The oldest of these ancient outcroppings have been found in Greenland, but they occur in all the continents in at least some area of the land mass. There are large outcroppings of ancient rocks in most of western Australia, the Ontario region of Canada, eastern South America, parts of India, Scandinavia, and Africa, and elsewhere. Not all of these are 3 or 4 billion years old, but

---

* The use of a superscript number at the right, called an *exponent,* provides a useful means of writing numbers, especially those that are large or awkward. When the exponent is positive it indicates the number of digits to precede a decimal point. For example, the number 1×10^9 indicates that there are nine zeros after the 1 and before the decimal point in the whole number, or 1,000,000,000. The number 4.6×10^9 is transcribed as 4,600,000,000 because there are nine digits after the last whole number (4) and before the decimal point. In this number the decimal is moved nine places to the right, after the whole number. It is a common practice to eliminate the number preceding the multiplication sign ($\times$) when that number is 1; thus 10^6 is understood to be 1×10^6 or 10 to the sixth power, or the number 1 million. When the exponent is negative, it indicates the number of zeroes *after* the decimal. For example, 5×10^{-3} is equivalent to one five-thousandth or 0.005; and 10^{-6} is equivalent to one-millionth, or 0.000001.

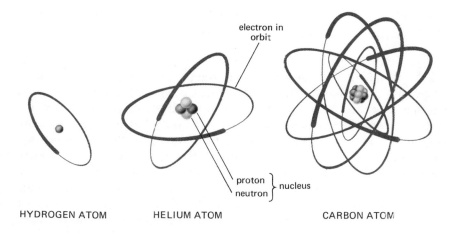

electron in
orbit

proton
neutron } nucleus

HYDROGEN ATOM   HELIUM ATOM   CARBON ATOM

Figure I.2   The atom fundamentally consists of a compact nucleus surrounded
by orbiting electrons. The hydrogen nucleus has one proton and no neutrons,
but all other atoms include both of these particle types. The helium nucleus
(alpha-particle) includes two protons and two neutrons; the carbon nucleus has
six protons and six neutrons. The electrons circle in orbits under the influence
of their attraction by the positive charge of the protons in the atomic nucleus.

many are 2 to 2.5 billion years of age. Since much of the Earth's crust has
changed considerably and since erosion has occurred everywhere, the Earth
must be older than 4 billion years. Rocks of this age are presumed to be
the oldest *remaining* crustal material, the more ancient layers having been
lost by geological and climatic processes, which cause disintegration. Esti-
mates of greater age for the Earth have come from dating studies of meteorites that have landed on the planet at various times in its history. The
meteorites are 4 billion 600 million (4.6×10^9) years old, no matter which
sample is analyzed or which decay series is used for the determinations.
The consistency of these age determinations from independent lines of evidence is interpreted to mean that the age of the entire solar system is at least
of this value. Since we believe that the whole solar system was formed at
about the same time, the Earth also must be at least 4.6×10^9 years old.
Meteorites are considered to be parts of the solid materials of the solar
system and a representative sampling of it.

Various problems are encountered in dating studies using these decay
series. An element of error is inherent in each method for various reasons,
and samples of such materials can be dated only if they happen to contain
these rare radioactive atoms. In cases where volatile matter such as the
gases argon or helium are necessary for determinations, losses of these
materials may prevent analysis or distort the calculations. Regardless of the

difficulties, no other methods are available to date solid matter of such antiquity.

CARBON-ISOTOPE METHOD FOR DATING ORGANIC MATERIALS

In addition to the commoner isotope of carbon, ^{12}C, the atmosphere also contains ^{14}C. These two isotopes occur in the atmosphere in a constant ratio, and ^{14}C decays to ^{12}C with a half-life of about 5700 years. When a sample of organic material, such as seeds, wooden objects, mummy wrappings, and so on, is to be analyzed, it is volatilized first to convert the carbon to carbon dioxide, a gaseous molecular form. Once CO_2 is formed, the proportion of $^{14}CO_2$ to $^{12}CO_2$ is determined and compared with the proportion of these two forms in the present-day atmosphere. The age of the sample is calculated from this proportion and from the known half-life of ^{14}C decay to ^{12}C. A constancy is assumed because all organic material contains carbon and the ultimate source of all carbon is the CO_2 of the atmosphere, which is incorporated into organic molecules by the process of photosynthesis in green plants. This organic material is utilized by all life as a consequence of the food chain in which animals derive their food from plants. There are inherent problems in this method as in most others, so inaccuracies may be encountered. But, more importantly, because the isotope has such a short half-life, this particular method cannot be used for materials of any great age. Materials older than 30,000 years or so are generally unsuitable for carbon-isotope measurements, because so little ^{14}C remains for analysis.

OTHER METHODS FOR DATING

Dating of rock formations sometimes can be made on the basis of the fossils embedded in these rocks, the thickness of the rock strata, and the age of strata adjacent to those under study, and by other methods that rely upon geological data. For recent deposits from postglacial times it sometimes is possible to date materials by analyzing the pollen samples that are present. The pollen identifies the trees that grew in the area, which in turn indicate the climate of the time. These data are pooled to determine the age of the sample under study.

Methods have been developed to estimate the age of the universe from observations of the color and relative brightness of stars. The color is an index of temperature; studying color together with brightness makes it possible to theorize on the sorts of atomic reactions occurring in the star. These items of information are interpreted in terms of the stage of evolution of the star; hence its age on the basis of predictable reaction sequences. If all the assumptions are accepted, then such calculations have indicated 10 to 12 billion years for the age of the oldest stars observed. This brings us very close to the edge of the observable universe, beyond which we cannot see and therefore cannot know, on the basis of current physical laws and principles. The various methods for estimating the age of stars and of the universe are somewhat controversial, and unequivocal absolute values cannot be given at present.

Some Suggested Readings

Dingle, H., "Cosmology and science." *Scientific American* 195 (September, 1956), 224.

Jastrow, R., *Red Giants and White Dwarfs*. New York, Harper & Row, 1967.

Seaborg, G., and J. L. Bloom, "The synthetic elements: IV." *Scientific American* 220 (April, 1969), 56.

CHAPTER 2

COSMOLOGY

The universe, also known as the cosmos or metagalaxy, includes all existing matter in space. The branch of astronomy that deals with the general structure and the origin of the universe is known as cosmology. The chief structured components of the universe are the masses of matter in galaxies and star clusters, and the dust and gas of interstellar areas. There are estimated to be 1 billion (10^9) galaxies in the universe and an average of 100 billion (10^{11}) stars in each of these galaxies. If we accept these figures, then there are about 10^{20} stars (10^9 galaxies $\times$ 10^{11} stars per galaxy $= 10^{20}$ stars) distributed throughout the universe.

Just as stars occur in clusters or in the larger aggregation of the galaxy, the galaxies themselves tend to occur in clusters. Our own galaxy is one of 17 in a cluster estimated to have a diameter of 2 million light-years. Outside this cluster our nearest neighbor galaxy is a spiral galaxy 8 million light years away. Galaxies occur in various shapes; according to some astronomers and cosmologists these shapes may indicate an evolutionary sequence of change of an individual galaxy.

About 99 percent of matter in the universe is in the form of elemental hydrogen and helium, and only 1 percent of the abundance is comprised of all the remaining elements, which are all heavier than hydrogen and helium (Fig. 2.1). The calculated ratio for the universe as a whole is 10,000 H:500 He:1 heavy atom, but there are local variations of this proportion. For example, cosmic rays consist of about 87 percent protons (nuclei of hydrogen atoms), 12 percent alpha particles (nuclei of helium atoms), and 1 percent heavy-atom nuclei. Unlike this distribution in deep space, our own sun contains a higher proportion of He:H and thus is considered to be younger than some other components in space. On the basis of the assumed origin of matter and the universe, and the known transmutations of matter, it is possible to derive an index of relative age based upon the proportions of atoms in a particular sample.

From spectrographic analyses of elements in different parts of the observable universe, it is clear that a common set of elements make up all matter. No unusual or unique elements have been found anywhere, but the elements of the periodic table do occur in various proportions in different regions of the universe. This is not surprising in view of the observed differences in aggregations of matter in space, that is, of stars and gases and dust clouds and the dilute distribution of elements in interstellar space. Although spectrographic analysis can be made of only the surface of a star and not of its interior composition, about 70 of the known elements have been detected in our sun, a typical middle-aged star.

Observations of the stars provide the basis for proposed sequences of evolutionary changes in the universe. The luminosity or relative brightness of a star, called its magnitude, indicates a transfer of energy across space. This energy is produced by the atomic reactions occurring on the surface and in the interior of the star. The energy we see radiating into space represents the energy that is lost in the reactions, which originally was stored in atoms and molecules in the star. According to the equation $E = mc^2$, there must be a reduction in mass at the source of the radiation of energy, since these are interconvertible. Such a reduction with time represents an evolutionary change, as we define such phenomena.

The temperature, size, and mean density of stars also are suggestive of evolution. Surface temperatures generally vary from 3,000°C to 30,000°C, as interpreted from the variation in color, from cool reddish through yellow and green to hot bluish stars. Stars may be one-tenth the size of our sun or a million times greater in volume. Mean densities vary from those of the collapsed and degenerate white dwarfs, which are 1000 times denser than water, to the supergiant red stars with a density one-millionth that of water.

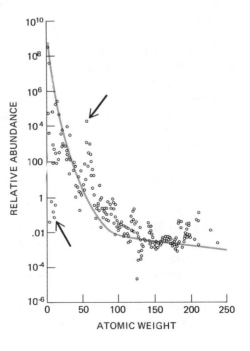

Figure 2.1   The relative abundance of the elements is plotted on a logarithmic (exponential) scale according to atomic weight. There is an excess of members of the iron group (near atomic weight 56) and far less of the elements near atomic weight 10, such as lithium, beryllium, and boron, when compared with the theoretically expected abundance curve (solid line).

Evolutionary sequences have been constructed using such data. The density of the supergiant red stars is similar to the density of matter in interstellar space, while white dwarfs may represent stars that are very close to extinction.

Origin of the Universe

Two principal theories of universe origin are acknowledged today. The "Big Bang" Theory for an exploding universe first was proposed in the 1930s by the Belgian Father Lemaître and later was expanded mathematically by George Gamow, the American physicist. They postulated that a primeval mass of matter (finite according to Lemaître, infinite according to Gamow) of great density underwent a radioactive explosion about 10 billion years ago. The universe began to expand at that time and has been expanding ever since. From the mathematical calculations made on the basis of energy, mass, and the prevailing conditions at the beginning, Gamow proposed that matter was dispersed homogeneously at first but that gravitational forces increased and led to condensations. Gas clouds formed at first and developed into protogalaxies. Later, as condensations continued, stars formed and galaxies developed more or less as we see them today in the universe.

A central issue of this theory of the exploding universe concerns the formation of the elements. It is postulated that only protons, neutrons, and electrons existed at the beginning. But after about 5 minutes the universe cooled enough to permit interactions between neutrons and protons, to form deuterium. All the remaining elements formed during a period of about 30 minutes, by the process of neutron capture. Because of the temperatures calculated to have existed at that time, Gamow considered that thermonuclear reactions would not have continued for longer than 30 minutes. Also, free neutrons are extremely unstable, and any that were not used up in element formation would have decayed to protons and electrons by that time. Thirty minutes is not an unreasonable interval for these events, although it may seem to be at first glance. If we consider the analogy of an atomic explosion and its resulting fission products, then we can think of microseconds relative to years as a proportion that is comparable to minutes relative to billions of years. During experiments performed on atolls of the Pacific Ocean in recent years, it was observed that one microsecond of thermonuclear reaction time generated some elemental fission products that were still present at the explosion site at least 3 years later. This ratio of 1 microsecond to 3 years is the equivalent of 30 minutes to 5 billion years.

Although there are lines of evidence in support of this theory, some objections were raised quite soon after its proposal. There is clear evidence from experiments using particle accelerators that neutron capture cannot proceed from helium, with its atomic weight of 4, to the next unit of the series with an atomic weight of 5. Helium is very stable; when bombarded with neutrons, as experimental evidence reveals, ^{5}He breaks down almost immediately to ^{4}He. A similar situation is known to be true for building elements beyond beryllium (^{8}Be) by neutron capture. Even more significantly, we know now that element synthesis occurs in stars. In view of these data, Gamow modified the exploding universe theory to state that all the lighter elements were synthesized during the first 30 minutes of universe existence but that heavier elements were synthesized later in the hot interiors of stars.

The second major theory of universe origin was proposed in 1948 and later years by Bondi, Gold, and Hoyle. According to the Steady State Theory, the universe is infinite in time and space, with no beginning and no end. As matter disappears from our view at the limits of the observable universe, newly created matter replaces what is "lost." This process of renewal is assumed to occur at a constant rate throughout space. While masses of matter such as stars and galaxies do undergo changes with time, the overall large-scale features of the universe are unchanging. That is, if all matter in the universe were to be distributed uniformly throughout space, its properties could theoretically be considered to be unchanging.

According to the mathematical premises presented by Hoyle, the overall stability of the universe is a function of the continuous creation of new matter, which just compensates for the separating effects of universe expansion. Space does not become emptier each year, and it would require the creation of only one new atom of hydrogen every thousand years to maintain a constant density of matter in the universe as a whole. The Steady State Theory assumes that the fundamental laws of physics and chemistry are constant and have been the same throughout time. The exploding universe theory, on the other hand, requires that matter behave differently before and after the initial explosion. These particular alternatives cannot be resolved by any tests known at the present time.

A variety of evidence is available upon which to judge these two opposing theories of universe origin. Both theories include the expansion of the universe, for which there is ample evidence. The analysis of particular elements using either reflecting telescopes or radio telescopes reveals specific absorption bands in the spectrum of radiation of the element under study. A "red shift" in the spectrum is observed, indicating that matter is moving away from us (the observer). The shift toward the red end of the spectrum indicates an increase in the wavelengths of radiation of particular absorption bands. The farther away the matter happens to be, the greater is the shift of spectral bands toward the region of longer wavelengths in the red portion. The expansion of the universe is indicated also by determinations of the speed of travel of stars and galaxies throughout the universe.

There is a linear correlation between the increased speed of travel and the increasing distance of the object from the observer. This correlation has been expressed as Hubble's Law (Fig. 2.2). Thus, at a distance of 10^6 light-years the expansion speed is 30 miles per second, at 10^8 light-years it is 3000 miles per second, at 10^9 light-years it is 30,000 miles per second, and so forth. The phenomenon of the red shift is similar to the Döppler effect observed in the behavior of sound. As sound moves away from the listener, the signal becomes lower in pitch with increasing distance because the sound waves lengthen. The usual familiar changes in the sound of an approaching or receding train whistle serve as an example of this phenomenon. If the sound waves are approaching, then they are shorter and we hear a higher pitched sound; if moving away from us, the lengthening sound waves reach our ears as lower in pitch.

The synthesis of elements in the stars is an acknowledged phenomenon, and provides one of the strongest lines of support for the Steady State Theory. However, the exploding universe theory can accommodate this occurrence also. Either of these two theories receives support from the observations, and we cannot choose the correct alternative on this basis any more than on the premises of universe expansion.

One of the more promising approaches to deciding between these alternative theories is that of determining the density of matter in space, particularly in relation to the distribution of the galaxies in the universe. According to the Big Bang Theory, one would predict an irregular distribution of galaxies, since supposedly they arose by random condensations of gas clouds. But there should be a greater concentration of galaxies farther away than nearer to us (the observer) because all matter began to move away after the initial explosion of the infinite primeval mass, and at the same time. After these intervening billions of years since the explosion, according to predictions based on the theory, we should detect a significant

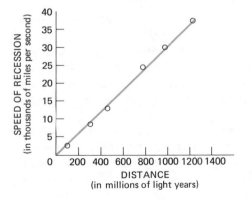

Figure 2.2 The linear relation between speed of recession and distance from the observer has been plotted for various galaxies (shown as dots); the observed points lie on a straight line, as predicted by Hubble's Law.

increase in density of galaxy distribution the farther we go out in space. The Steady State Theory predicts a relatively uniform dispersal of galaxies (matter) throughout the universe, with no particular difference in distribution farther away or nearer to the observer. If matter in the form of hydrogen atoms is distributed uniformly and created uniformly throughout space, then condensations of matter in the form of galaxies and galaxy clusters also should be distributed more or less uniformly in the observable universe.

There is a great deal of astronomical data from both reflecting- and radio-telescope studies concerning sources of radiation in the universe, whether the radiation is that of wavelengths of the visible light spectrum or the very long radio waves, among others. With reflecting telescopes galaxies can be seen clearly and confidently out to a distance of only about 1 or 2 billion light-years. That is, the light we observe now from these bodies first began their travel toward us 1 or 2 billion years ago, and we just now are seeing this radiation long after the journey began. Some of these galaxies may have become extinct in the meantime, but we would not know this until we compiled billions of years of observation, which seems unlikely to happen. The farther out into the universe we can perceive matter, the more of the history of the universe should be available to us. Signals from radio sources may be interpreted as coming from galaxies many billions of light-years away, but few comparative data are available to permit us to identify such sources unequivocally as being at such great distances. Similar controversial interpretations surround the evidence derived from studies of quasars, an abbreviated term for quasi-stellar radio sources. Quasars are extremely luminous objects, much more luminous than ordinary galaxies. Originally they were considered to be starlike objects that also showed strong radio emissions, but this restriction no longer applies. If we were sure that quasars are 8 or 9 billion light-years distant in the universe, we could study their distribution in relation to the different predictions made by the two major theories of universe origin. But there is no general agreement that quasars are so distant, because it is not known with certainty whether their demonstrable red shift is a consequence of universe expansion or of some other phenomenon. If their considerable red shifts are a function of expansion, then the increasingly higher concentration of these starlike bodies 8 to 9 billion light-years away from us would serve as supporting evidence for the concept of an exploding universe rather than one in a steady state. Unfortunately, the peculiar combination of characteristics of quasars leads to different interpretations of their relative nearness to us as observers; therefore, such data are not necessarily relevant to universe origin. At present, it is not possible to examine galaxy distribution for unequivocal evidence for either theory of universe development. As new evidence is collected, favor shifts from one theory to the other. At the moment, the theory of the exploding universe is considered to be the most likely of the two explanations for the origin of the cosmos in which we exist.

STELLAR SYNTHESIS OF THE ELEMENTS

An important component of both theories of universe origin concerns element synthesis in stars, and much of the evidence is compatible with either theory. It is important for us to consider element synthesis in relation to formation and evolution of stars, in view of the fact that atoms combine to form molecules and molecules of particular kinds led ultimately to the origin of life on the Earth. A logical progression of thought can be derived from such considerations, and we will pursue this progression in the context of the discussion of the origin of life constituents.

According to the theoretical sequence of birth, development, and death of a star, the evolutionary progression is assumed to begin with a cold, dilute, turbulent tract of gas and dust in interstellar space. Gravitational forces lead to condensations into stars, and, as contraction continues to occur, the interiors of the stars become very dense and very hot. When the core of a star reaches a temperature of about 5 million degrees, protons move with sufficient energy to fuse into deuterons upon colliding. Further collisions between the deuterons, which are composed of two protons, and other free protons lead to the formation of helium. From laboratory experiments it is known that ^{3}He does not react further with free protons, but that two ^{3}He nuclei will fuse to form ^{4}He plus free protons. By such proton–proton fusions (Fig. 2.3), the net result is the conversion of four hydrogen nuclei to one atom of helium with a concomitant release of enormous amounts of energy. The proton–proton fusion sequence is the chain-reaction phenomenon principally responsible for the energy source in most stars. Other energy-releasing atomic reactions occur rarely in the life

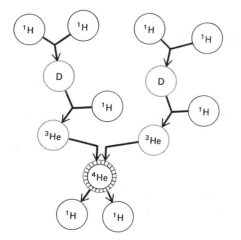

Figure 2.3 The proton-proton fusion sequence in stars leads to the release of enormous amounts of energy during the formation of helium from hydrogen nuclei. Two protons (^{1}H) fuse to form a deuteron (D), which interacts further with free protons to yield helium-3 (^{3}He). The combination of two helium nuclei produces stable helium-4 (^{4}He) plus two free protons. Four protons are thus consumed for each helium nucleus that is produced.

of a star. These studies and others conducted by Hans Bethe, a recent Nobel laureate in physics, have provided us with much of our understanding of stellar reactions and energetics.

As proton–proton fusions continue, an internal core of helium develops gradually in the star and increases in size with time. But as the store of hydrogen fuel is used up, the core of the star begins to cool. During cooling, gravitational forces predominate and contraction takes place. With contraction, the temperature of the star's core begins to rise again. The rise in temperature is relatively rapid in the core, and these internal temperatures cause heating of the outer hydrogen envelope of the star, leading to a considerable expansion of this outer mantle. The expanded surface of the star then radiates cooler (redder) light and the star enters the "red giant" stage of its evolution.

Such a red giant star has a core of helium at the very high temperature of about 100 million degrees. Under these conditions, pairs of ^{4}He nuclei fuse to form atoms of ^{8}Be (beryllium). The atoms of ^{8}Be are produced almost as rapidly as they distintegrate, but a small amount of ^{8}Be is believed to remain at all times under these conditions. These occasional ^{8}Be nuclei will fuse with ^{4}He, under the presumed conditions of the red giant core, to form an atom of ^{12}C (carbon) in an initial excited (high energy) state. When the ^{12}C nucleus becomes stable, the energy equivalent of 7.6 mev (million electron volts) photons is released. Such specific energy release has been observed, thus verifying the occurrence of such atomic reactions in these stars.

These reactions explain the increasing occurrence of ^{4}He and ^{12}C but do not account for the synthesis of the very rare elements, lithium, boron, and beryllium, nor for their relatively low abundance in the universe (see Fig. 2.1). It has been postulated that these three elements are generated as products of atomic fissions during rare stellar explosions, not as a consequence of regular reaction processes in stars. The ^{8}Be isotope described in the ^{12}C synthesis sequence does not accumulate, because such nuclei fuse with helium to form carbon. The curve of element abundance in the universe more closely approximates the predictions made in the Steady State Theory than those for an exploding universe and serves to provide strong support for the Steady State concept.

The synthesis of elements heavier than carbon occurs by various processes (Fig. 2.4), including helium capture. By this process involving fusions with ^{4}He, ^{12}C is built up in successive steps to ^{16}O (oxygen), ^{20}Ne (neon), and ^{24}Mg (magnesium). As ^{4}He is used up in such element formation, the helium core of the star first cools and then contracts because of the cooling. The contraction in turn leads to a rise in core temperature. Such temperatures may be high enough to permit the synthesis in some stars of various members of the silicon group of elements (^{28}Si and others) from nuclei of carbon, oxygen, and neon atoms. When the core temperature reaches

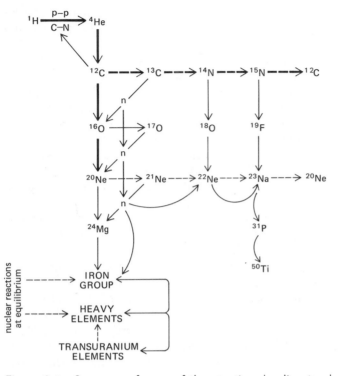

Figure 2.4   Summary of some of the reactions leading to element synthesis in stars. Reactions include those in which fusions occur with protons (→), helium nuclei (↓), and neutrons (– n –), among others such as the catalytic processes (- →) of the Carbon–Nitrogen and Neon–Sodium Cycles. (Adapted from "Formation of Elements in Stars," M. Burbidge and G. Burbidge, *Science*, Vol. 128, pp. 387–398, 22 August 1958, with permission of the American Association for the Advancement of Science.)

about 5 billion degrees, it is assumed that no further fusions will occur. Shortly before this temperature is achieved, members of the iron and nickel (^{56}Fe, ^{59}Ni, and others) group will be formed from magnesium and other lighter elements present. The elements of the iron–nickel group are among the most stable of all those known. These elements absorb rather than release energy in reactions, so iron cannot serve as fuel for further chain reactions. The observed overabundance of iron in the universe has been suggested by Hoyle to be a function of its accumulation in primeval stars and its subsequent release into space upon star explosion (supernovation).

While proton–proton fusions lead to lighter elements in younger stars, and members of the heavier groups (such as those to which iron and nickel belong) are believed to be produced by various reactions involving carbon and helium and a series of intermediate elements in red giant stars, still heavier elements are believed to be generated in older second-generation stars. Such second-generation stars are presumed to be born from the dust and gas released during star explosions, upon condensation under the influence of gravitational forces. The materials released from a nova (star explosion involving only the outer mantle) or a supernova (explosion of entire star) would be of various sorts, distributed in various parts of the universe where such events chanced to take place. New second-generation stars would contain elements that happened to be present in the areas of condensation, including elements from the iron and nickel group.

An important source of energy in the universe is found in second-generation stars, which are very hot (at least 18 million degrees). In addition to the proton–proton fusion sequence that occurs, one helium atom is produced from four hydrogen nuclei in the carbon–nitrogen cycle, first proposed in 1939 by Hans Bethe. This synthesis of helium is catalyzed by carbon and leads to the production of enormous amounts of energy. The conversion of four atoms of hydrogen to one of helium results in the loss of 4 million tons of mass per second in a star. The conversion of this mass to energy is acknowledged according to the Einstein equation $E = mc^2$. Our own sun is assumed to derive only 1 percent of its energy supply through the carbon–nitrogen cycle and 99 percent from proton–proton fusions.

The proposed sequence of events involves the incorporation of ^1H in a series of steps, which also leads to the ejection of positrons (e^{1+}) from atomic nuclei and the generation of energy. The final step in the cycle is one of atomic fission, by which ^{12}C is regenerated and a new atom of ^4He is created. Intermediate steps involve the intervention of various isotopic forms of nitrogen, carbon, and oxygen.

A variety of other thermonuclear reactions occur in the interior of a hot star. The fusion of a proton with ^{16}O leads to the isotope ^{17}O, and a similar reaction produces ^{21}Ne isotopes from ^{20}Ne. Together with ^{13}C, the elements ^{17}O and ^{21}Ne play a very important role in further element synthesis in the core of a red giant star, where there is a large supply of helium. When ^{13}C, ^{17}O, and ^{21}Ne interact with ^4He, unstable nuclei are produced, which emit neutrons. This steady supply of free neutrons in the hot core of the star is utilized by heavier elements that readily capture neutrons. The formation of elements with atomic weights between those of the iron–nickel group and the lead–bismuth series is assumed to be a function of neutron capture in such older red giants. When these red giants explode, the neutron-capture process occurs faster than the rate of decay of the newly forming elements. In this way it is believed that the heaviest known elements are synthesized during star explosions. Spectral analyses of supernovae have revealed the presence of very heavy elements that are not known to exist in other locations in the universe. The unstable element technetium has been detected; because this element has a very brief half-life it must have formed more recently than the star in which it was observed. That is, the element must have been synthesized in that star. Similarly, the presence of the element californium with an atomic weight of 254 has been observed in spectral data from some supernovae. The only other places where this element has been found have been at sites of thermonuclear test explosions on Earth and as products of reactions in particle accelerators such as bevatrons. Such data are considered evidence in support of stellar syntheses of heavy elements, as well as the more frequently generated lighter elements.

All the data and calculations from stellar syntheses fit the observed abundance curve of the elements in the universe. The calculations of the time required for this element abundance distribution have been on the order of 1 to 2 billion years. The universe clearly is older than this, and even our own solar system is almost 5 billion years old. A sufficient amount of time obviously has elapsed to explain the abundance of the elements on either the exploding universe or the steady state theories. Such information, therefore, does not permit us to choose between these two alternatives either.

Origin of the Solar System and the Planet Earth

The most widely accepted theory of the origin of the solar system was first proposed in the eighteenth century by Kant and LaPlace and expanded and modernized in the present century by Fred Whipple and others. There was considerable resistance to the acceptance of this hypothesis for a long time because it was assumed that interstellar space was a vacuum, devoid of all matter. Current interpretations based in general on the Dust Cloud Hypothesis assume a rather elaborate sequence of events and specific changes in temperatures. The ideas are attractive for many reasons, but especially because similar formations may be assumed to have occurred in all parts of the universe at many different times. According to the Dust Cloud Hypothesis, we would expect solar systems to be frequent components of the universe and not at all rare phenomena. Alternative theories, such as those premised on collisions between stars or extrusion of matter from one star as a result of tidal actions set up when another star passed close by, would lead us to expect solar systems to be very rare occurrences. Moreover, solar systems formed by such alternative events would be very different from the type deduced according to the conditions set forth in the Dust Cloud Hypothesis or any of its variations. Of course, we have little evidence to support our belief that solar systems abound in the universe. But there is evidence of a solar system for a relatively nearby star, and it is unlikely that two solar systems would have been formed so close together in the vastness of space if any kind of rare accident was responsible for their existence.

According to the Dust Cloud Hypothesis, the first stage involves the formation of a star (sun) from a solar dust cloud as a result of condensations of gases and dust in a dilute dispersion, under the influence of gravitational forces. We assume that the forerunner dust cloud was similar to those that are observed today. Thus the temperature must have been close to absolute zero, and various elements must have been present, although hydrogen and helium would have predominated. In addition to various elements in atomic form, it is assumed that molecules also would have been present. Particular combinations of atoms into molecules would have included H_2, CH_4, and various inert gases; NH_3, H_2O, iron oxides, and compounds containing various atomic linkages including C—C, N—N, N—C, and C—O. It is further assumed that the condensing protostar would have been surrounded by a large rotating cloud of dust and noncondensible gases such as hydrogen and helium, but that no particle aggregates or large bodies would have been present in that cloud.

Planetesimal formation would have signaled the second stage of solar system development. Formation of the star would have led to atomic reactions and consequent heating of those parts of the cloud closest to the radiations from the new star. The coagulating effect of liquids and damp bodies (similar in principle to the greater ease of making snowballs from wet snow than from drier snow) would lead to the formation of larger and larger aggregates of particles within the clouds of gases. The newly forming planetesimals would incorporate such nonvolatile compounds as iron oxides and sulfides, solid ammonia and water, hydrated minerals, and other materials. Gases also would be associated with these aggregates. It is considered possible that temperatures within the evolving dust cloud would have been similar to those that exist today in various parts of the solar system.

Protoplanet formation would occur during the third stage, under conditions of higher temperatures. Due to gravitational forces, protoplanetary gases would be compressed and thus cause increased temperatures (above 2000°C) in the central region. As the planetesimals and protoplanets passed through this heated gaseous medium, some of the gases would escape, many aggregates would be volatilized, and a variety of chemical changes would occur involving iron and silicon compounds. It is assumed that the amount of iron would have become relatively greater as volatilizations occurred and gases escaped. The rarity of many gases in our atmosphere is assumed to be due to their escape during this stage of planet development. These events also explain, at least in part, the relatively high content of iron for our own planet.

During the fourth stage when the final accumulation of the Earth occurred, it is believed that the temperature dropped to about 0°C as energy radiated away. If this is correct, then the Earth formed when temperatures were typical of its present condition. Small amounts of gases would have been retained owing to the presence of weak gravitational fields at these lower temperatures, so the suggestion actually is an attempt to explain conditions that characterize the Earth as we know it now. For this same reason, it is suggested that the high content of iron and silicon was merely a function of chance distribution in particles from which the Earth was formed during the sequence of stages just described. Various types of meteorites also show mineral compositions that could be due to such chance particle presence in the particular dust cloud from which our solar system formed.

The completion of the Earth and the moon represent the fifth stage of formation. The lithosphere (solid matter) of the planet contains slightly more silicate than iron compounds in the crust. Although we are not certain, there is geological evidence that the core of the Earth was formed later in Earth's history, long after the crust had solidified. The atmosphere (gaseous matter) was highly reducing in character (i.e., little or no molecular oxygen was present), containing the original hydrogen, methane, hydrogen sulfide, ammonia, and water vapors. But because of the temperatures assumed to have existed at the time, it is considered most likely that a great deal of hydrogen gas escaped, whereas heavier gases were retained by gravitational forces. Such a reducing atmosphere is a crucial feature of the theories proposed to explain the later origin of life on Earth. The liquid component or hydrosphere of the Earth is believed to have been about 10 percent of the present amount. More water is believed to have been added later, from hydrated silicates and especially from condensed water of the Earth's interior.

The moon probably was completed about the same time as Earth, according to the most widely accepted ideas. Expeditions to the moon may produce evidence of the satellite's age if appropriate rock samples can be secured for analysis, either by humans or by robot mechanisms. Since the Earth is presumed to be the same age as the rest of the solar system, having formed from a common solar dust cloud, the moon also should be 4.6 billion years old if it was completed at the same time. We may know very soon whether or not this is the case.

Some Suggested Readings

Burbidge, M., and G. Burbidge, "Formation of elements in the stars." *Science* 128 (1958), 387.

Fowler, W. A., "The origin of the elements." *Scientific American* 195 (September, 1956), 82.

Gamow, G., "The evolutionary universe." *Scientific American* 195 (September, 1956), 136.

Hoyle, F., "The steady-state universe." *Scientific American* 195 (September, 1956), 157.

Jastrow, R., *Red Giants and White Dwarfs.* New York, Harper & Row, 1967.

Sandage, A. R., "The red-shift." *Scientific American* 195 (September, 1956), 170.

Schmidt, M., and F. Bello, "The evolution of quasars." *Scientific American* 224 (May, 1971), 54.

CHAPTER 3

THE ORIGIN
OF
LIFE ON THE EARTH

If life did originate sometime during the Earth's history, then we must explain the development of those complex organic molecules that are unique to life as we know it and the sequences of events that led from a chaos of chemicals to an organized living system. From our previous considerations it seems clear that simple kinds of molecules were present on the primeval Earth, such as water, ammonia, and methane, among others. In addition we expect that more complex organic compounds also occurred, in view of evidence of their presence in stars, on other planets of our solar system, and in the varied forms of meteorites that have fallen on the planet. If we accept the stipulation that each of these locales is unable to support life, then the chemical constituents we observe in such objects must have been synthesized abiotically. Similar abiotic syntheses presumably could have occurred during the early history of the Earth, long before the appearance of life forms, since there would be few significant differences for syntheses of similar chemicals in various regions of the solar system.

Chemical
Evolution

The first comprehensive studies of abiogenic synthesis of organic compounds relative to the Earth's history were reported in 1922 by the Soviet biochemist A. I. Oparin. Many additional lines of evidence have been reported from his laboratory in the intervening 50 years, and these studies have provided a comprehensive framework for understanding the events that led ultimately to the first organisms on the Earth. A number of stipulations have been made concerning conditions that must have prevailed on the primeval Earth to permit the synthesis and accumulation of organic molecules. Among these are the reducing character of the atmosphere and the availability of ample sources of free energy for abiogenic syntheses. Because it is considered very unlikely that the atmosphere was oxidizing at that time, the existence of a reducing atmosphere on the primeval planet is an essential condition for the kinds of chemical reactions that have been studied and assumed to have occurred in ancient times. The transition to an oxidizing atmosphere containing levels of molecular oxygen similar to today's must have occurred billions of years after the Earth was formed.

Among the various sources of free energy most likely to serve for abiotic synthesis, solar radiation probably was the most significant. The longer ultraviolet wavelengths, between 300 and 400 nanometers, were especially important for the specific reactions critical to the origin of life. Electrical discharges may have been more common in primeval times, but we have little knowledge about the frequency of storms that may have characterized the Earth then. Thermal energy may have been a significant component, but if the Earth formed under conditions in which temperatures were much like those of today, then thermal energy would be an unlikely source for primeval abiotic reactions. Most reactions proceed slowly at lower temperatures, and reaction rates would not have been sufficiently rapid if the Earth's crust was relatively cool during its earlier history, as we believe it was. Similarly, energy derived from decay of radioactive elements probably was insignificant for primeval syntheses, although we cannot rule out this possibility entirely.

A number of experiments have been reported in which the design was such as to simulate general conditions on the primeval Earth and then to determine the patterns of chemical synthesis that would take place in such duplicated environments. Because the atmosphere lacked molecular oxygen, the available molecules would have occurred in a reduced form. Thus methane (CH_4) would serve as a carbon source, ammonia (NH_3) as a nitrogen source, and water (H_2O) as a source of oxygen atoms for more complex molecules. Some high-energy source, such as ultraviolet light, electrical discharge, or heat, would be applied to the system containing simple gaseous mixtures, and the endproducts would be isolated and identified after a suitable interval for the reactions to take place. Further requirements for the simulated system would include the maintenance of temperatures below 100°C, in accordance with accepted ideas about the Earth's formation and possible prevailing temperatures, and the exclusion of any life form by the usual precautions of sterilization of the components of the system. In one of the first experiments of this sort to be reported, Stanley Miller showed that a variety of simple organic molecules were produced, including amino acids. The significance and excitement of this finding was that amino acids are the building blocks of proteins, which are essential for life as we know it. The carbon–nitrogen linkages of amino acids would be crucial events for the eventual evolution of living forms, in which many kinds of proteins serve as functional and structural components of the organism. The control experimental system lacked the particular energy source, available as electrical discharge in Miller's first experiments, and infinitesimal numbers of organic molecules formed in the absence of energy.

Thus microorganisms could not have been responsible for the organic syntheses, since modern life would have produced molecules in the absence of
free energy in the system. The endproducts clearly were synthesized abiotically in the laboratory experiment and, by analogy, could also have been
produced under similar conditions on the primeval Earth.

Numerous experiments of various designs have been described since
1953. The general picture that has emerged provides substantial evidence
that a host of significant organic chemicals cou   ave been formed abiotically in primeval times. The more recent studies   ve concentrated on questions concerning the synthesis of nucleic acids a   energy-storing molecules
such as adenosine triphosphate (ATP). From   eral laboratories, including numerous reports by Cyril Ponnamperuma it seems clear that most of
the unique and significant molecules required for the origin of life could
have been synthesized abiotically and may have been present in reasonable
quantities in primeval times. As syntheses proceeded, organic molecules
accumulated in greater concentrations in the primeval seas and in greater
diversity as a result of condensations and polymerizations into larger molecules from smaller ones that were present.

Whereas there is general agreement concerning the presence of complex
organic compounds on the Earth, there is less agreement about the populations and organization of the reaction components and products of abiotic
syntheses. According to one extreme view, the entire primeval sea may
have been "metabolizing" as a single vast entity from which life forms later
were derived. Chemical evolution thus would be viewed as a function of
groups of chemical reactions in disorganized arrays and not ordered into
separated units or structures until much later. A more widely accepted view
is that the chemical reaction systems essentially were separated from one
another and from their general surroundings. In other words, some form of
morphological organization must have been present and some sort of
boundary must have delimited the reactions that occurred from the background environment. In this latter concept there is an acknowledgment of
the presence of order amid the increasingly chaotic or disordered surroundings. While increasing disorder (greater entropy) occurs in the universe as
a whole, according to the Second Law of Thermodynamics, the sequence of
events leading to life must have been such as to impose less disorder and,
therefore, greater probability for improvement, interactions, and complexity
of chemical reaction systems. If chemicals occurred willy-nilly in the
primeval seas, there would be no compelling rationale for the development
of increasing levels of organization and for the retention of interacting reactions in the same location. In considering the events that led to protobiont formation, we will accept the concept that there was increasing order
in delimited systems concomitant with thermodynamic increase in entropy
in the disarray of the surrounding primeval seas.

Evolution
of the Protobionts

Among the suggestions and experimental studies dealing with protobiont formation, the major ideas have been presented by A. I. Oparin, who favors coacervates, and by Sidney Fox, who leans toward the development of protein spheroids on the primitive Earth. In both cases the essential problem is one of understanding the development and increasing propensity for organization and coordination in time and in space. Life forms of all sorts share the common features of exquisite packaging and miniaturization of complex and interacting components within a delimited structure. The myriad chemical reactions are interrelated and coordinated *spatially* within the cell in modern forms, occurring in particular three-dimensional arrangements, and are regulated *temporally* such that all possible reactions do not occur all the time. This kind of organization also must have characterized the protobiont ancestral forms of the first living organisms on Earth. Our basic premise in every discussion of organic evolution is of descent with modifications. Therefore, the systems we study should have features related to the life that evolved ultimately and should permit us to see a chain of events that serves as a logical progression of relationships and modifications. If life forms have spatial and temporal organization, then we assume that similar conditions prevailed in the protobionts from which life arose eventually. If we accept these assumptions, we can proceed next to examine possible models by which such coordinated and organized improvements may have developed during the early history' of the Earth.

From Oparin's experiments and observations, we can consider the coacervate droplet as a model system for the primeval protobiont. An attractive aspect of this particular model is that it involves colloidal particles and water, two basic features of life forms. During the process of coacervation, colloidal particles join together in a common droplet as a function of mutual exclusion of some of the water molecules surrounding the separate particles. With the exclusion of some of the water molecules, the particles become parts of a single coalesced system, which now is sharply separated from the remaining water and other materials of the surroundings (Fig. 3.1). Many such coacervates may form in a common medium; complex coacervates also may form. Thus different kinds of particles come to be associated in the same droplet. There are several significant consequences of coacervate formation, relative to known traits of life. A boundary layer separates the droplet from the chaotic surroundings and provides a localized system in which interactions become possible without interference from the many chemical compounds and reactions of the environment. The coacervate may include any of a number of chemically different constituents in any number of combinations, providing a level of diversity from which

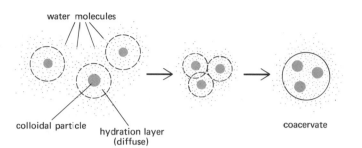

water molecules

colloidal particle   hydration layer
(diffuse)

coacervate

Figure 3.1   Diagram showing the process of coacervation. The exclusion of some of the water molecules from the hydration layer, and the delimitation of remainder of the layer, leads to a coacervate droplet, which contains one or more kinds of colloidal particles. The dashed or solid lines bounding the droplets represent the boundary layer separating the bound water within the droplet from the free water outside.

more successful combinations may come into existence by chance alone. Even though the seas may have contained relatively low concentrations of chemicals, coacervation would permit an increase in the concentration of chemicals in a confined area. Such a mechanism increases the probability for the coming together of related organic compounds or coparticipants in orderly chemical reactions. In a typical experiment, a solution of the protein gelatin containing one part protein per 100,000 parts of water may change after coacervation so that 95 percent or more of the protein from the solution becomes concentrated locally in the coacervate droplets. This mechanism simultaneously provides for the concentration of reactants from dilute seas or solutions and for their separation from the surroundings via containment in structural entities.

Coacervate droplets often show some level of internal structure, as when vacuoles form within the systems. Internal structure is a universal feature of modern cells and provides a physical basis for spatial separation of chemical reaction components within a coacervate. The boundary layer between the droplet and its surroundings also confers the advantage of selective entry and exit of molecules and thus displays surface properties to be expected in a predecessor of life. Finally, increasing complexity may be generated by the coalescence of simple coacervates to form complex droplets containing a variety of organic molecules and enhanced internal compartmentalization.

From all these characteristics, it seems that coacervate droplets provide many of the required features to be expected in a protobiont ancestor of primeval life. However, there are some difficulties in accepting this system in its entirety as a prelife candidate. Coacervates tend to be rather transient, forming and disintegrating rapidly under many conditions. Such fragile droplets may not have persisted long enough to incorporate improvements and modifications, especially under wide variations in acidity and alkalinity of the medium in which they form. The coacervates are stable within relatively narrow limits of acidity and alkalinity of the environment, and they may have enjoyed extremely brief episodes in many regions of the primeval seas. Despite these reservations, the coacervate does provide a reasonable model for protocell formation whereby increasingly ordered and complex protobionts could have developed prior to the appearance of the first life forms.

Another model for protocell formation is the protein spheroid or microspherule as proposed by Fox and others. When dilute solutions containing 18 of the common amino acids are exposed to boiling water temperatures, small spherical structures appear in great quantities. These microspheres are rather stable, as shown by their ability to withstand high centrifugal forces and chemical fixation or exposure to harsh conditions. Preparations of such structures for study with the electron microscope do not degrade, and when thin sections of these spheroids are examined there is clear evidence of the presence of a boundary or outside layering. These spheroids change shape in salt solutions, which indicates an osmotic effect due to selective penetration of chemicals between spheroids and surroundings. In some ways these polymerized amino acids are unnatural forms of proteins, particularly in the way peptide bonds form from linkages of carbon and nitrogen atoms in the molecules. Such spheroids display a negative response in the Gram stain reaction when they are constituted of acidic amino acids; they are Gram-positive when basic amino acids predominate in the polymers. These traits of the proteinoid structures fulfill a number of desirable features expected of protobionts. One serious defect, however, is the lack of diversity among such spheroid populations. Considering the low probabilities for the appearance of a life form among protobionts, the chances seem even slimmer for relatively homogeneous systems such as protein spheroids. It is entirely possible that spheroids did form on the primeval Earth, as would be probable physically and chemically in various phase-separation phenomena in which bubbles and droplets appear. Whether the proteinoid was a significant step leading to further development among protobionts, however, is the basic question, a question that cannot be answered unequivocally. Neither system is entirely suitable, nor can either system be discarded altogether as a possible predecessor for life forms.

Aside from the fundamental requirement for morphogenicity as a framework for organization in space and time, protobionts with the potential for improvement would also be required to prevent an increase in entropy in the system. Organization must be maintained and enhanced, and disorder must be minimized or abolished in a protobiont for it to persist long enough to incorporate improvements in managing its affairs amid a swarm of competing reactions and a finite supply of raw materials for existence. Unless there is a flow of energy and matter between the protobiont and its surroundings, the system will either remain static or distintegrate. Of the various exchange systems that have been described, the *open system* is the one characterized as undergoing a constant exchange of both energy and matter with the environment. Since living organisms are open systems, and since entropy would increase in any system that did not maintain a supply of free energy, we expect the successful protobionts to have been open systems. Such protobionts would be more likely to display an interval of survival that would be longer than intervals of formation or disintegration. The open system would be a fairly static arrangement, however, unless it was capable of achieving stable states, that is, of accommodating a variety of different equilibrium conditions in response to the changing external environment. An open system that can assume a stable condition under one set of chemical reactions and switch to other stable conditions according to various possible sets of reactions is called a *stationary open system.* Life forms are stationary open systems that demonstrate many stable states. Such stable states occur at different times and in great variety in the development and existence of a cell or organism, and they generally exist simultaneously under temporal regulation and coordination in even the simplest cell.

A particular stationary state is maintained as a function of an exchange of energy and matter during chemical reactions and is particularly related to rates of these reactions and rates of diffusion of substances in and out of the system. If the rates of reactions or of diffusion change, new levels of organic compounds occur and a new stable state is established. An infinite number of stationary states is possible, depending principally upon the rates of the chemical reactions that occur. This arrangement permits considerable flexibility for development and increases the diversity of protobiont types that may have formed in primeval times. If the reactions are coordinated rather than randomly related, then the system is capable of indefinite levels of adjustment to changing conditions.

What sort of advantage would have to be incorporated into the protobiont stationary open system to permit a continuing development in the direction of life? One favorable, if not essential, feature would be self-renewal, since such systems would persist longer than others that lacked this ability. If some systems happened to include reactions that permitted repair or replenishment of components, they would be more likely to persist than would systems unable to restore their constituents to appropriate levels. Simply because nonrenewable entities ultimately would degrade as entropy increased, or disintegrate for any reason, the self-replenishing forms would last longer. It follows, therefore, that persistent protobionts would become more numerous by accumulation in the environment. This is not to be confused with differential reproduction, which is a unique quality of life and the basis for changing proportions of life forms according to natural selection. In the case of the protobionts we are dealing with a relative change in frequency because some types persist and are joined by other newly formed persisting entities. Protobionts that existed fleetingly would not become predominant components of primeval populations. Since there is no particular advantage to persistence per se, we must ask how the capacity for persistence could lead to further improvement. The simplest answer is that a persistent form is more likely to incorporate additional or more favorable reactions because it remains in existence longer than its neighbors. There would be more chance for modifications in an entity that persisted, but only chance would determine whether such modifications actually improved the system. There certainly would be incorporations that reduced the advantages of some protobionts, as well as others that enhanced activities. But only those forms with improved capacities for existence and for interactions of an increasingly dynamic nature would continue to exist and, perhaps, predominate in some location in the primeval seas. One particularly advantageous capacity would be the ability to increase in mass, that is, to grow in size. Any change that led to a net gain in free energy might also permit growth, because additional syntheses could occur and more organic molecules might be retained in the system. The most probable kind of added reaction would be one coordinated with those already present in a protobiont. This would be an efficient improvement, one more likely to permit the utilization of energy and matter and thus lead to retention rather than loss of protobiont constituents. Energy not channeled into reactions or stored in chemical molecules would probably be lost as heat. In a typical experiment reported by Oparin, a coacervate droplet containing an enzyme protein and the carbohydrate starch was immersed in a solution of the energy-containing precursor glucose-1-phosphate. When the latter penetrated the droplet it was converted to starch, resulting in growth of the coacervate. If an additional enzyme protein was present in a similar coacervate, then the starch decomposed to maltose, which was excreted into the medium, thus establishing a flow of metabolites according to the following scheme:

$$\text{sugar}_1 - \textcircled{P} \rightarrow \text{sugar}_1 - \textcircled{P} \xrightarrow{\text{enzyme}_1} P_i + \text{starch} \xrightarrow{\text{enzyme}_2} \text{sugar}_2 \rightarrow \text{sugar}_2$$
$$\downarrow$$
$$P_i$$

Similar models have been created and studied, with every sort of biologically significant molecule included in one combination or another. From many such experiments, Oparin showed clearly that the systems with the capacity to grow were those containing closely coordinated reaction components, whereas coacervates with more disorganized constituents usually did not increase in mass, or disintegrated. This pattern of protobiont "selection" would be a prerequisite to continued improvements leading ultimately to some combination of materials and events that constituted a living system.

As protobionts assumed greater chemical and structural complexity, additional advantages would accrue to systems capable of the most rapid rate of reaction. Such systems would be more likely to obtain a steady supply of energy and chemical building blocks in competition with an increasing number and variety of droplets in the primeval seas. There would be even greater advantage in the competition for raw materials for those protobionts that incorporated cyclic reaction series. Such prelife forms would then be capable of regenerating some of the required reaction components and would not be dependent entirely on an external supply of each ingredient for each chemical reaction.

Stepwise additions of increasingly advantageous reactions would be one plausible pathway to improvement, but not the only one. If larger protobionts broke up into smaller units, reaction systems would be separated, but they could recombine in different ways upon coalescence of smaller units into larger new units. Such new combinations might yield efficiently interacting reaction systems more rapidly than single additive improvements could. Since coacervates readily undergo coalescence to produce complex droplets with internal structure and spatial organization, there is some basis for giving greater preference to this model for the primeval prelife form than to other forms already mentioned.

REQUIREMENTS FOR LIFE

Although the development of structures containing chemical reactions is a necessary step toward life, protobionts are considered to lack the ability to reproduce a continuing series of descendants similar to themselves. Each protobiont represents a unique experiment in chemical evolution and has no mechanism, as far as we know from comparisons with modern life forms, for perpetuating its own kind or for increasing the frequency of its unique features among the diverse populations on the primeval Earth. Living systems contain *thermodynamic, kinetic,* and *template* components, whereas prelife forms are believed to have lacked at least one of these. During chemical evolution one or more kinds of thermodynamic molecules must have been incorporated into successful protobionts, since chemical syntheses require the expenditure of free energy. Various energy-rich molecules function as thermodynamic or energetic components in modern life forms, but adenosine triphosphate is one of the most ubiquitous of these. The free energy potential contained in the thermodynamic molecule must be mobilized for synthesis, and the kinetic component accomplishes this function. Various kinds of catalysts act as kinetic components that mobilize the energy contained in molecules such as ATP, but the unique catalysts of life are enzyme proteins. Thus a protobiont may contain starting materials or endproducts of chemical reactions, but these reactions can proceed only if free energy is available in usable form. Catalysts also modulate rate of reaction, usually by lowering the energy of activation barrier (Fig. 3.2). In the example presented above, glucose-1-phosphate contains energy stored in its chemical bonds, which is made available for synthesis by virtue of enzyme action in cleaving the phosphate group from the sugar molecule. The enzyme also permits a rapid rate of reaction, far greater than the rate of spontaneous chemical change in its absence. In this particular example glucose-1-phosphate is both the raw material that is converted to the starch endproduct and the source of the free energy for the reaction. But in many reactions the energetic component is a different molecule from the starting chemicals involved in endproduct synthesis. In any case the principles are the same whether or not the chemical precursor is the vehicle providing the energy for the reaction.

According to laboratory experiments, thermodynamic molecules could have been synthesized abiotically and thus been readily available for protobiont chemistry. Whether or not enzymes were present and incorporated into prelife forms is a very different matter. Enzymes are specific catalysts and ordinarily cannot moderate a variety of reactions. Also, enzyme proteins are comprised of unique kinds and arrangements of amino acids, which in turn are translations of genetic information stored in nucleic acid template molecules. It is possible of course that some abiotically synthesized proteins possessed catalytic activity, but it is much less likely that precisely the right kind of enzyme just happened to become associated with exactly the right sets of chemical reactants. Probabilities for such coincidences would be very low, albeit not beyond the realm of possibility. However,

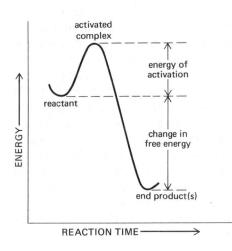

Figure 3.2   The energetics of a chemical reaction are depicted to show that the conversion of a reactant to one or more endproducts requires overcoming the barrier of the energy of activation. Activation energy must be supplied before the reaction can proceed and the potential or free energy of the reactant molecule can be released during the formation of the endproducts. Catalysts form a temporary activated complex with the reactant, thus lowering the energy barrier and modulating the rate of the chemical reaction.

because many kinds of inorganic compounds possess catalytic activity, though usually of a nonspecific nature, the general assumption is that primeval catalysts probably were inorganic rather than organic compounds in the beginning.

In living systems the template, thermodynamic, and kinetic components provide the framework for the coupled reactions and the many levels of regulation and coordination of activities that ensure maintenance, growth, and reproduction of the organism. A major unanswered question on the origin of life concerns the sequence of steps by which all three essential components came together in such a way as to act in concert and permit the reproduction of the whole pattern in successive generations. It is agreed that all components eventually came to exist in the first *eubionts*—that is, life forms—but there is little agreement on the sequence of these crucial events.

In simplistic terms, the question is whether the protobionts were "metabolically alive" or "genetically alive" before they became eubionts containing genetic and metabolic systems acting in related pathways in a common system. It is agreed that proteins and nucleic acids are essential for life as we know it. But which came "first" in the evolutionary sequence? If the protobionts evolved as protein-containing systems and later incorporated nucleic acid templates, then how did the correct informational templates come to be associated with the precise group of proteins translated from such information? Considering the precision of linear sequences of units in the template which are translated into the linear sequence of amino acids in each protein (Fig. 3.3), the coincidence is staggering to contemplate. On the other hand, if the protobiont contained the nucleic acid templates

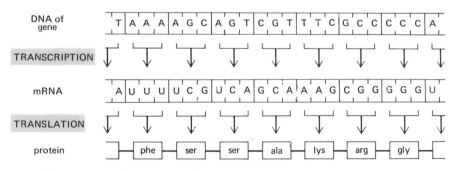

Figure 3.3   The codons (triplets of nucleotides) in the linear sequence of DNA are transcribed into messenger RNA intermediates and become translated into a specific linear sequence of amino acids, which comprise a polypeptide or protein molecule. The portion of the gene and polypeptide shown here represent accurate relationships between particular possible DNA codons and the amino acids *phenylalanine, serine, alanine, lysine, arginine,* and *glycine,* from among the 20 possible kinds of amino acids included in the genetic code (see Fig. 3.7).

and no related proteins, then how can we explain the occurrence of chemical reactions and the synthesis of molecules in a stationary open system? The nucleic acid molecules have four major genetic properties: (1) information storage, (2) mutation potential, (3) information transfer, and (4) precise replication. But only the first two properties are functions of the nucleic acid molecule itself, that is, of its intrinsic arrangements of atoms and subunits. Information transfer from the template to the synthesis system and replication of the molecule both require the participation of specific enzymes and energy sources, at the least (Fig. 3.4). Apparently neither

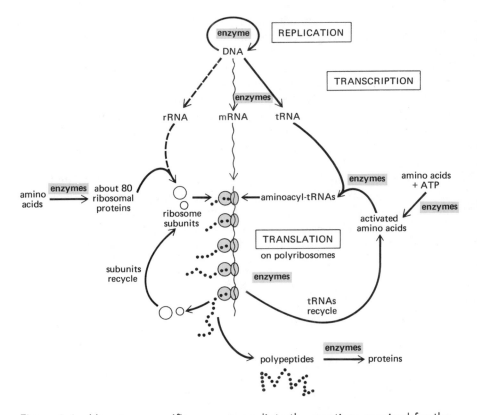

Figure 3.4   Numerous specific enzymes mediate the reactions required for the replication of DNA, transcription of DNA information to RNA molecules, formation of ribosomal proteins which combine with ribosomal RNA to form the particulate ribosomes, and the activation of amino acids which then combine with transfer RNA molecules under the direction of many aminoacyl-tRNA synthetases. The synthesis of a specific polypeptide or protein molecule according to information encoded in the DNA template thus requires the intervention of well over 100 different enzymes prior to the ultimate translation, each of these intervening enzymes also being products of gene action.

protein nor nucleic acid alone fulfills all the requirements for a prior system to which the other component can be added later, to constitute life. Because of these difficulties in understanding how either kind of molecule could have existed alone and then come together in very precise and snugly fitting ways, various studies have been conducted to determine whether either or both of these protobiont models could be considered *evolvable,* that is, capable of undergoing continued improvement in the absence of the other part of the team. As an example of one approach to the solution of the dilemma of "which came first," we will consider a series of experiments designed to answer the question, Can template replication occur without the participation of enzymes? Clearly, this question usually is asked by those who favor the gene-first–enzyme-later hypothesis.

When all four bases of ribonucleic acid (RNA) or deoxyribonucleic acid (DNA) are present in organic solvents, adenine pairs with uracil or thymine, and guanine pairs preferentially with cytosine (Fig. 3.5). Thus, the pairing of complementary bases is a function of the bases themselves and not of interactions involving other kinds of molecules, too. Since complementary base-pairing is a required feature of the duplex DNA molecules and of the varied interactions between DNA and RNA and between RNA

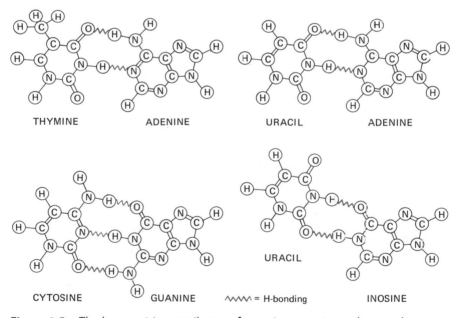

Figure 3.5   The base-pairing attributes of cytosine, guanine, adenine, thymine, uracil, and inosine. The first three kinds of nitrogenous bases occur in both DNA and RNA, whereas uracil is a conventional RNA component and inosine (a deamination product of adenine) occurs in some kinds of RNA molecules. Thymine is unique to the DNA macromolecule.

and RNA (Fig. 3.6) in the genetic apparatus, this line of information is important in freeing the gene of dependence on enzymes for this trait, as well as for the features of information storage and mutation potential referred to earlier in this discussion. In seeking further information about nucleic acid capabilities, it was found that stable helix formation occurred when polynucleotides and mononucleotides were placed together in dilute solutions. Thus, polyuracil served to orient adenosine-5'-monophosphate, and polycytosine acted similarly with guanosine-5'-monophosphate. The polynucleotide chains were able to orient appropriate mononucleotides and thus acted as templates even though enzymes were absent from the experimental system. Although the data provide evidence that nucleic acids can function independently of proteins, at least at the level of orienting mono-

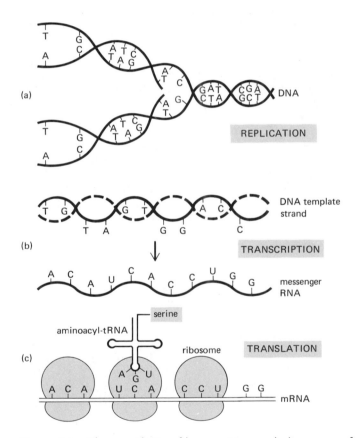

Figure 3.6   The specificity of base-pairing underlies many of the interactions between nucleic acid molecules, such as (a) replication of DNA, (b) transcription of RNA from DNA, and (c) the codon–anticodon recognition system of messenger RNA and transfer RNA during translation on the ribosomes.

nucleotides, the evidence was inadequate for at least two reasons. First, the particular linkages that formed as the mononucleotides joined to form oligomers on either template were unnatural, since they do not occur in modern nucleic acids; second, we must know whether these newly formed oligomers can serve as templates in turn and provide a basis for continuity through descendant generations.

If we have a nucleic acid template comprised of adenine and cytosine in an alternating sequence, A—C—A—C—A—C—A—, and U—G—U—G—U—G—U— forms as its complement in a double helix configuration, then we have the beginnings of a replication system. But the system cannot be perpetuated unless the complement in turn can act as a template and lead to the continued production of A—C—A—C—A—C—A—. If this does not occur, then the consequence would be the accumulation of complementary molecules and no additional templates of the original type, except as these might be produced by chance polymerizations and thus increase in frequency at random. Such a distortion would not fulfill the requirements for a perpetuating system based on the generation of templates and complements in an equivalent progression. In one experiment of this sort it was reported that polyuracil could act as a template to orient monomers of adenylic acid, which formed polyadenylate oligomers. But polyadenylate did not act as a template in turn for polyuracil formation. In this one case, at least, the system lacked requirements for perpetuation of templates.

Unfortunately there has been very little definitive study of this problem, so a detailed pursuit of answers would be unprofitable at the moment. Regardless of the specific disagreements and the gaps in our understanding of the sequences of events involved in the transition from chemical to biological evolution, there is general agreement on a number of points. It is generally conceded that life did originate on this planet rather than having been transplanted to the Earth from elsewhere in the solar system or outer space. The appearance of life on Earth undoubtedly was a consequence of the same natural phenomena of physics and chemistry as occur today on Earth and elsewhere in the universe. Given these generalities, we may deduce that life can originate anywhere in the universe at any time that suitable conditions prevail, and it may have done so many times. Indeed, simple life forms may be developing at this moment on the Earth in places that permit an appropriate sequence of events. But such newly created life would be difficult to detect with any methods now available to us and probably would be destroyed by existing life forms which utilize such materials for their own growth and reproduction. Clearly, protobiont accumulation on the primeval Earth was made possible by the absence of life forms at the time. With the appearance of the first eubionts, any protobionts would have led a precarious existence at best and may have been eradicated very shortly after life originated on our planet.

Origin
of the Genetic Code

The genetic code provides a basis for our understanding of the means by which information is stored in compact form in the nucleic acid molecules, the way this information is translated accurately into protein products of gene action, the recognition of codons and anticodons in the translation process, and the punctuation marks for initiation and termination of the messages translated. There are three general characteristics of the genetic code: (1) Only four kinds of nitrogenous bases are utilized in the messenger RNA transcripts of the template information, these being uracil, adenine, guanine, and cytosine (U, A, G, C); (2) codons are composed of triplets of bases in all possible permutations, so that four kinds of bases grouped in threes permit 64 different codons ($4^3 = 64$); and (3) only 20 of the naturally occurring amino acids are incorporated into the genetic code for all life forms today. To approach the problem of the origin of the code, it would be profitable to examine some of its features for clues to the solution to the problem. Since the first substantial breakthrough reported in 1961 by M. W. Nirenberg and J. H. Matthaei, showing that the RNA codon UUU translated as the amino acid phenylalanine, a great deal has been learned about the code (Fig. 3.7).

Upon examination of the codons and their amino acid translation counterparts, it is immediately apparent that the 20 amino acids are not distributed randomly among the 64 codons. In particular: (1) Codons that end in U or C but have the same first two bases—that is, xyU or xyC—always code for the same amino acid, as in the case of the codons UUU and UUC for phenylalanine, UAU and UAC for tyrosine, CAU and CAC for histidine, and so on. (2) Codons xyA and xyG often code for the same amino acid, as with AAA and AAG for lysine, GAA and GAG for glutamic acid, and others, except for the presumably more recent amino acids methionine (AUG) and tryptophan (UGG), each of which has only one codon. (3) In half the 16 groupings the general codon $xy*$ codes for a single amino acid, where the third base in the sequence (at the 3' end) as represented by $*$ may be any of the four possible bases—for example, GGU, GGC, GGA, and GGG code for glycine; ACU, ACC, ACA, and ACG code for threonine; UCU, UCC, UCA, and UCG code for serine. (4) If the first two bases are any combination of Gs and Cs (GG, GC, CC, CG) then the four codons sharing the same initial doublet all code for the same amino acid, indicating that the meanings of these codons are independent of the third base in the sequence. In fact, this is true for all codons with C in the middle position—for example, GCU, GCC, GCA, and GCG all code for alanine. And (5) in most cases all the codons for a particular amino acid start with the same doublet, with three exceptions all of which are examples of amino acids coded by six different codons—

Second Nucleotide

Third Nucleotide →

First Nucleotide ↓ (RNA codons in italics)

First nucleotide A or U

Third	Second: A or U	Second: G or C	Second: T or A	Second: C or G
A or U	AAA / *UUU* — phenylalanine	AGA / *UCU* — serine	ATA / *UAU* — tyrosine	ACA / *UGU* — cysteine
G or C	AAG / *UUC* — phenylalanine	AGG / *UCC* — serine	ATG / *UAC* — tyrosine	ACG / *UGC* — cysteine
T or A	AAT / *UUA* — leucine	AGT / *UCA* — serine	ATT / *UAA* — "stop"	ACT / *UGA* — "stop"
C or G	AAC / *UUG* — leucine	AGC / *UCG* — serine	ATC / *UAG* — "stop"	ACC / *UGG* — tryptophan

First nucleotide G or C

Third	Second: A or U	Second: G or C	Second: T or A	Second: C or G
A or U	GAA / *CUU* — leucine	GGA / *CCU* — proline	GTA / *CAU* — histidine	GCA / *CGU* — arginine
G or C	GAG / *CUC* — leucine	GGG / *CCC* — proline	GTG / *CAC* — histidine	GCG / *CGC* — arginine
T or A	GAT / *CUA* — leucine	GGT / *CCA* — proline	GTT / *CAA* — glutamine	GCT / *CGA* — arginine
C or G	GAC / *CUG* — leucine	GGC / *CCG* — proline	GTC / *CAG* — glutamine	GCC / *CGG* — arginine

First nucleotide T or A

Third	Second: A or U	Second: G or C	Second: T or A	Second: C or G
A or U	TAA / *AUU* — isoleucine	TGA / *ACU* — threonine	TTA / *AAU* — asparagine	TCA / *AGU* — serine
G or C	TAG / *AUC* — isoleucine	TGG / *ACC* — threonine	TTG / *AAC* — asparagine	TCG / *AGC* — serine
T or A	TAT / *AUA* — isoleucine	TGT / *ACA* — threonine	TTT / *AAA* — lysine	TCT / *AGA* — arginine
C or G	TAC / *AUG* — methionine	TGC / *ACG* — threonine	TTC / *AAG* — lysine	TCC / *AGG* — arginine

First nucleotide C or G

Third	Second: A or U	Second: G or C	Second: T or A	Second: C or G
A or U	CAA / *GUU* — valine	CGA / *GCU* — alanine	CTA / *GAU* — aspartic acid	CCA / *GGU* — glycine
G or C	CAG / *GUC* — valine	CGG / *GCC* — alanine	CTG / *GAC* — aspartic acid	CCG / *GGC* — glycine
T or A	CAT / *GUA* — valine	CGT / *GCA* — alanine	CTT / *GAA* — glutamic acid	CCT / *GGA* — glycine
C or G	CAC / *GUG* — valine	CGC / *GCG* — alanine	CTC / *GAG* — glutamic acid	CCC / *GGG* — glycine

Figure 3.7 The genetic code. The nucleotide triplets of the genic DNA are shown in italics, the RNA codons in boldface type, and the amino acids or punctuations in lower-case. Note the redundancy of the code except for the amino acids methionine and tryptophan.

namely, leucine by CU* + UUA + UUG; serine by UC* + AGU + AGC; and arginine by CG* + AGG + AGU. In addition to these interesting correlations there is the general impression that "related" amino acids have related codons, at least to some extent. Thus all codons with U in the middle position code for hydrophobic amino acids, and the basic and acidic amino acids appear to be grouped together in the standard distribution chart (Figure 3.7, at the bottom right).

Considering these special features of the code, F. H. C. Crick has suggested some possible interpretations in relation to the origin of the genetic code in primeval times. In all probability the primitive code was based on a triplet codon pattern rather than on doublets or even singlet bases; that is, the code did not evolve by the addition of bases to make the eventual triplet seen today. As we know from genetic analysis the genetic message is read linearly along a reading frame, and the addition or deletion of a single base in the sequence throws off the entire reading frame to produce a defective protein (Fig. 3.8). Similarly, if evolution had proceeded for a time with smaller codons and later expanded to a triplet codon system, the reading frame would become totally distorted and would produce lethal defects. If we accept the premises that the triplet codon existed from the beginning and that the reading frame was preserved through later modifications, then we can examine some possible models for the primitive triplet codon pattern. It may have been a system in which only the first two bases formed the code words and the third base functioned stereochemically or in some other noncoding fashion. If only the first two bases specified particular amino acids, then the dictionary would have contained only 16 codons based on four kinds of bases in all possible doublet combinations (4^2) and the third base would not have participated in codon specificity. Even if the triplet codons contained only two *kinds* of bases instead of the four kinds in modern nucleic acids, and all the three bases contributed to codon specificity in a triplet, this still would have permitted up to eight different amino acid translations ($2^3 = 8$). There is evidence in support of the notion that some kinds of amino acids are more ancient constituents of proteins. In those proteins considered to be ancient in origin, amino acids such as glycine, alanine, serine, and aspartic acid occur with high frequency. The amino acids methionine and tryptophan are found less often and usually in proteins considered to have evolved more recently. Such lines of evidence provide indications that a smaller dictionary may have existed in primeval times, that is, that there were fewer codons and fewer kinds of amino acids specified for proteins by these triplets. Crick considered it unlikely that nonsense or punctuation codons existed in any great numbers in the primeval code, principally because this would have led to more gaps in the proteins than could be tolerated. Only three termination codons exist in the modern genetic code, namely, UAA, UAG, and UGA.

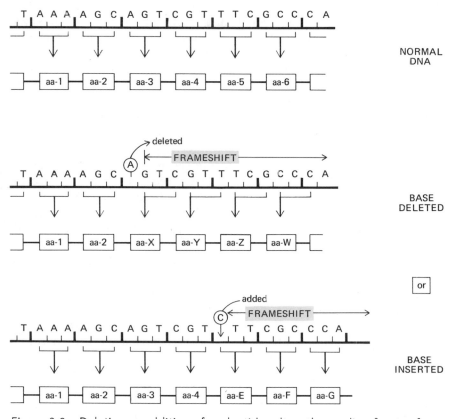

Figure 3.8   Deletion or addition of nucleotides alters the reading frame of the genetic message, leading to changed protein translation products.

On the basis of a triplet code, how many kinds of bases might there have been in the primitive nucleic acids? For the molecule to be informational there must have been more than one kind of base or the code would have been limited to only one amino acid. Furthermore, since complementarity operates in the genetic apparatus, we would expect at least two kinds of bases, for example, A + U or T, or G + C. Crick suggested that adenine and inosine would be likely candidates for primeval nucleic acids if only two kinds existed at the beginning. Adenine probably was one of the commonest nitrogenous bases in the primeval seas, according to experimental studies of abiotic syntheses. Inosine can be converted to adenine by deamination and would be expected to have coding properties similar to guanine. Although it is true that codons made up of As and Gs code for amino acids that are believed to be ancient, there really is little other evidence in support of Crick's suggestion. But the evolution of a four-base code from a two-base code could occur without upsetting the reading frame, and that is the crucial point. Continuity would be possible because newer messages could include original triplet codons and triplets added during the evolution of protein synthesis.

Fragmentary as the evidence may be, we can suggest a possible sequence of events by which the code arrived at its present level of evolutionary development. Whether the codons were permutations of two or four kinds of bases, there must have been fewer amino acids included in the early genetic code. It is most probable that these same few amino acids were translated from almost all the primeval triplets, with very few codons specifying punctuations, as is the case in modern genetic messages. A simple set of transfer RNA (tRNA) species would have been sufficient for a recognition system that included fewer amino acids and codons, and few activating enzymes would have been required for aminoacylations. Thus, the early proteins may have been quite simple in amino acid composition, and perhaps rather crudely made. As new amino acids evolved there would have been concomitant expansion of the transfer RNA and activating enzyme complements, and increasing precision of the recognition system in the successful primeval organisms. As new codons and transfer RNAs were added, it is possible that the anticodon regions of tRNAs would recognize some existing and some new codons as well. In this fashion, the present ambiguity of the codon—anticodon system may have become established. According to Crick's "wobble" concept, for which there is experimental evidence, the base at the 5' end of the anticodon is less confined spatially than the other two bases in the group. This permits the base to form hydrogen bonds with any of several kinds of bases at the 3' end of the codon. Thus if G occupies the wobble position in the anticodon, it can pair either with U or C in the complementary codon; and if U occupies the 5' end of the anticodon, then it may recognize either A or G at the 3' end of the codon. There is greater restriction for C and A because each of these recognizes only one particular base at the 3' end of a codon, these being G and U, respectively. If inosine is in the wobble position of the anticodon it may form hydrogen bonds with A, U, or C at the proper codon site. Inosine arises in modern transfer RNAs by enzymatic modifications of adenine, but it may have been an original ingredient of the codon–anticodon recognition system in the earliest nucleic acids, as mentioned above. The demonstration of different hydrogen bonding capabilities of the 5' base in the anticodon provides a reasonable basis for explaining the observed fact that modern genetic systems contain fewer transfer RNA species than codons. Similarly, it provides an acceptable explanation for increases in the codon dictionary and the various components that interact during message translation, as a function of base-pairing recognition arrangements. Of course, as the code expanded we would expect new amino acids to be related to existing amino acids if the resulting polypeptides were to be functional and stable molecules. We also would expect new amino acids to confer some selective advantage on at least one protein into which these were incorporated; otherwise there would be no compelling basis for code expansion according to accepted premises of evolutionary theory.

A number of other interesting features of the genetic code deserve comment; among these we might consider the remarkable fact that the code is essentially universal in modern organisms. Regardless of the source of the cellular extracts—whether microbial, plant, or animal—a particular messenger RNA will stimulate the incorporation of a particular amino acid into a polypeptide chain. Thus polyuracil guides polyphenylalanine formation, polycytosine promotes proline incorporation, and polyadenine leads to polylysine molecules in all systems that have been examined. In more stringent tests of code universality it has been shown that many specific triplet codons will promote the incorporation of the same amino acids in an in vitro system, whether the transfer RNAs of the protein synthesizing system were derived from bacterial, amphibian, or mammalian cells. It is reasonable to expect this degree of invariability of the code if we consider the effects of mutations that might alter it. For example, if a mutation modified the tRNA that specified phenylalanine sequences (UUU) so that these tRNAs attached to serine (UCU) sequences instead, then such a mutant would synthesize proteins with few if any phenylalanine residues because the altered tRNAs would have inserted serine in place of the other amino acid. Such a mutant probably would prove to be lethal, since one or more of its essential proteins would be defective. There would be strong selection pressure against mutations of this sort, and the code would continue to be relatively invariant throughout evolution.

Such arguments are more persuasive when considering the restrictions imposed on random changes once the genetic codes had evolved toward greater complexity both of the codon dictionary and of the amino acids that were specified for increasing numbers and kinds of proteins. If we ask about the primeval code from which the modern complements were derived, it is even more difficult to rationalize the sequence of events by which it became established. Two general theories have been proposed to explain the observed universality of the genetic code. According to the "frozen accident" theory, the original allocation of codons was a matter of chance, but as life forms became more complex the code remained constant because of selection pressures to maintain systems that included the fewest lethal or detrimental effects. On the basis of the "stereochemical" theory, for which there is no particular evidence either, the similarity of the codes in life forms is due to the similarity of chemical requirements—in particular, to the stringencies imposed upon the system because of stereochemical relations between the amino acid and the codon. The idea is testable, and for this reason at least is attractive to some investigators. The frozen accident theory cannot be tested experimentally with methods now available.

In this discussion we have made the assumption that the primeval genetic materials were essentially the same as those we know today, particularly that DNA existed as the informational template. It is entirely possible that the early genetic systems were based upon RNA rather than DNA and that the DNA apparatus was incorporated later into continuing ancestral lines. We know that RNA viruses exist in many variations, so RNA can have an informational and a template function. For some of these RNA viruses, however, recent evidence has revealed the presence of polymerase enzymes which direct the synthesis of DNA templates from the RNA molecules introduced by the virus during infection. These newly formed DNA templates then continue to guide the formation of new viral RNA strands that later are incorporated into the mature virus particles at the end of the infection period. Thus even in an apparent RNA-only genetic system there is the intervention of DNA and its decisive template function for the viral cycle to be completed.

Many other kinds of alternatives have been proposed for the primeval genetic system, but progress has been limited because we understand only a part of the complex picture. For example, we know that modern ribosomes are composed of RNA and protein molecules in very specific combinations and arrangements. Perhaps the ancestral ribosomes were made only of RNA and stabilizing proteins were added later in evolution. We cannot answer many important questions now, but there is good reason to believe that approaches will become possible as we learn more about the many levels of interaction involved in the storage, transfer, and translation of genetic information into the myriad proteins in modern cells and organisms.

Some Suggested Readings

Calvin, M., *Chemical Evolution.* New York, Oxford University Press, 1969.

Crick, F. H. C., "The origin of the genetic code." *J. Molecular Biology* 38 (1968), 367.

Jukes, T. H., *Molecules and Evolution.* New York, Columbia University Press, 1966.

Kenyon, D. H., and G. Steinman, *Biochemical Predestination.* New York, McGraw-Hill, 1969.

Marshall, R. C., C. T. Caskey, and M. W. Nirenberg, "Fine structure of RNA codewords recognized by bacterial, amphibian, and mammalian transfer RNA." *Science* 155 (1967), 820.

Nirenberg, M. W., "Genetic code: II." *Scientific American* 208 (March, 1967), 80.

Oparin, A. I., *The Origin of Life on the Earth.* 3rd ed. New York, Academic Press, 1957.

Oparin, A. I., *Genesis and Evolutionary Development of Life.* New York, Academic Press, 1968.

Oparin, A. I., "Routes for the origin of the first forms of life." *Sub-Cellular Biochemistry* 1 (1971), 75.

Woese, C. R., *The Genetic Code: The Molecular Basis for Genetic Expression.* New York, Harper & Row, 1967, pp. 179–195.

CHAPTER 4

FURTHER EVOLUTION OF LIFE FORMS

If we examine the spectrum of life forms in existence today, we see an astonishing variety of sizes, shapes, and life styles. Underlying this incredible diversity is a multitude of metabolic capacities, specified by genetic information encoded in the DNA molecules. From the fossil record and from comparative biochemistry and physiology of living species, we know that some metabolic pathways have been in existence for a long period of evolutionary history while others have become greatly modified or disappeared all together. Any single chemical reaction usually is quite simple and, under appropriate conditions, may occur in a test tube as well as in the living cell. If we focus on the chemistry of life as sets of reactions occurring in cyclic and sequential steps, we become aware of the unity of metabolism as a consequence of the coordination and integration of related and intertwined chemical events. Metabolism is a complex of processes based upon regularity, sequence, and control, and therefore upon organization in time and space. It is this pheonemon of organization that distinguishes the chemistry of life from that of the nonliving world. Even the simplest virus contains information that leads to the synthesis of molecules in a temporal sequence and thence to spatial organization of the finished viral proteins and nucleic acids into a specified and distinctive three-dimensional structure. We know that enzymes mediate the chemical reactions of metabolism and that these unique catalysts of life also are ordered in time and in space within the organism. Furthermore, since genetic information specifies the manufacture or use of many kinds of proteins, including enzymes, we may reasonably infer that the evolutionary increases in numbers, kinds, and functional diversity of proteins depend basically upon increase and change in genes over the vast spans of evolutionary time. To grasp the significance of evolutionary improvement in primeval life we must examine the interaction and interdependence of genes and proteins in the display of metabolic variety.

Improvement
in the Catalytic Apparatus

A crucial feature of metabolism is the rate of chemical reactions, and enzymes not only modulate rates of reactions but also control the relation between rates of related reactions. The great diversity of organic compounds in living forms is due to the many combinations of a few kinds of reactions rather than to the occurrence of many different kinds of reactions themselves. Synthesis and degradation of chemicals often result from reaction sequences in which the same metabolites participate in a few basic types of reactions, which lead to a relatively large variety of endproducts via different combinations of metabolites and events (Fig. 4.1). Unlike the situation for organic molecules, inorganic compounds are less diverse but result from a greater selection of reaction types. Slow reactions are of little significance in living systems, whereas rapid reactions tend to occur in cycles and chains of steps. The reaction velocity is modulated by catalysts, and so we would expect that a major aspect of early evolutionary improvement would concern modifications of catalysts leading to enhanced reaction rates.

One principal means by which enzymes manage efficient catalysis is through their action in reducing the level of energy of activation for a

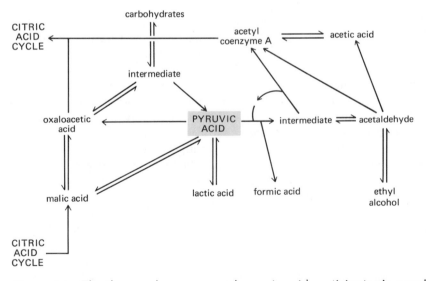

Figure 4.1 The three-carbon compound pyruvic acid participates in a variety of metabolic events, which lead to numerous endproducts, some of which are shown here. A few kinds of reactions yield many kinds of molecules in metabolic pathways that are interrelated and coordinated.

chemical reaction (see Fig. 3.2). A lowered barrier to the onset of a re-action permits the reaction to proceed more quickly. Another important feature of enzyme catalysis is their high specificity of action. Particular enzymes mediate particular reaction steps so that a selected group of reactions will occur in a cell, depending upon the enzymes that are present and active. In this way it is possible for many metabolites to be present but for only a few of these to participate in metabolism. The remarkable efficiency of enzyme action in the complex metabolism of the cell is therefore a function of the lowering of the energy of activation barrier and the consequent increase in reaction rate, plus the selectivity that permits parts of the total machinery to function at any one time.

We can imagine that primeval life was capable of a limited number of metabolic reactions, having a small number of genes encoding a finite amount of information. Some of these reactions probably were catalyzed by available components of a nonenzymatic type; others were perhaps mediated by the enzyme protein translations of primeval genes. Clearly, one line of improvement in primeval metabolism would be to incorporate more efficient catalysts into the organism, thus increasing the rates and specificities of chemical reactions. As an example of such evolutionary improvement, we will examine the chemical reaction in which hydrogen peroxide (H_2O_2) is broken down to molecules of water and oxygen:

$$H_2O_2 \xrightarrow{\text{catalyst}} H_2O + \tfrac{1}{2}O_2$$

One catalyst that can influence this reaction is iron, which was present in abundance on the primeval Earth. In its ferric (Fe^{3+}) or ferrous (Fe^{2+}) forms, iron acts as a catalyst for many kinds of reactions, only one of which is the oxidation of hydrogen peroxide. Since Fe^{2+} was present in primeval times, it undoubtedly mediated the production of water and oxygen from hydrogen peroxide in early life forms. This reaction system was not very efficient, but the breakdown occurred more rapidly than is possible by spontaneous, noncatalyzed changes. When ferrous iron is incorporated into the organic porphyrin compound, the heme molecule is 1000 times more effective as a catalyst for this reaction than is the case for Fe^{2+} alone. Heme contains a porphyrin moiety (Fig. 4.2), and there is excellent evidence that porphyrins were an abundant product of abiotic syntheses and that they easily combine to produce more complex molecular forms. The addition of iron to a suitable porphyrin complex yields the molecule heme, which is a component of many compound molecules in modern life forms. Thus, any primeval life able to utilize heme rather than Fe^{2+} alone as the catalyst would have had an advantage in the ancient populations. Not only would the rate of the reaction have been faster, but the reaction also would have been mediated by a more specific catalyst. A more specific catalyst would be involved with fewer kinds of reactions, thus improving

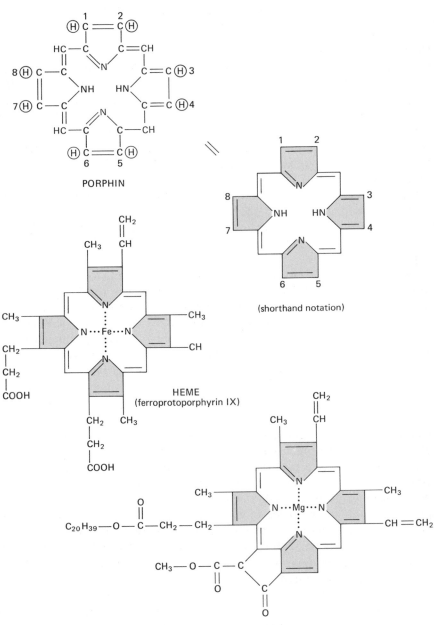

PORPHIN

(shorthand notation)

HEME
(ferroprotoporphyrin IX)

CHLOROPHYLL *a*

Figure 4.2   The porphin molecule may be chemically modified to a number of different porphyrin molecules, depending upon the substitutions at one or more of the eight carbon atoms indicated. The incorporation of a metal ion, such as iron in heme or magnesium in chlorophyll *a*, leads to further distinction and chemical possibilities in metabolism. The porphin nucleus of a molecule usually is indicated in an abbreviated form, as shown above.

efficiency. Less specific catalysts, on the other hand, would be involved in various chemical activities and be available for any one particular modification on fewer occasions. If this enhanced capacity was a heritable trait, then it would be transmitted to descendants, who themselves would be at an advantage relative to organisms that lacked this ability. Even if it was not inherited, organisms able to utilize more effective catalysts would be more likely to compete successfully for available raw materials required for growth and reproduction.

As increasingly effective catalysts led to improvements in metabolism, some members of the primeval populations must have become competent to synthesize proteins that could complex with compounds such as heme. These proteins, of course, would have been translations of new genetic information. When combined, the heme-protein molecules would have made up enzymes which would permit even greater metabolic efficiency than heme alone. Many modern enzymes consist of a protein conjugated to a prosthetic group, of which heme is one type. The particular enzyme catalase is just such an enzyme, and it sponsors the degradation of hydrogen peroxide 10 million times faster than is possible for heme lacking the protein. If this reaction were important for a life form, then those organisms able to synthesize catalase would enjoy a considerable selective advantage and might have come to predominate in some primeval populations.

In the proposed evolutionary progression just described, we envision an increasingly effective catalytic apparatus, which would confer advantages to some primeval organisms relative to others in a population. Each improvement is considered to result from a chance mutational change that would provide selective advantage; that is, there would be a greater probability for the improvement to be retained in the mutant and to be transmitted to descendant generations. Such improved organisms would be more likely to leave progeny, so we would expect later generations to include a higher frequency of the new metabolic capacity. It is also possible that such metabolically improved organisms would become the exclusive life form in some populations if less-advantaged neighbors left no descendants or produced fewer in each generation until no more remained In essence, we invoke the guiding principle of natural selection acting on mutant and nonmutant groups. Differential reproduction leads to increasing proportions of the more favorable types in a population, even if the original mutants were rare in the initial stages. Their frequency would increase by virtue of their leaving more descendants, which, like themselves, also left more descendants than would be produced by less-favored types. Since catalase is 10 million times more efficient than heme and 10 billion times more efficient than Fe^{2+} in processing hydrogen peroxide molecules, there would have been strong selection pressure to establish the mutant as the only or predominant type in a primeval population. This example can serve as a model to explain many other particular improvements in catalytic machinery. But the model is especially pertinent for biologically significant molecules which are comprised of protein conjugated to some kind of prosthetic group. It is interesting that modern cells contain various functional molecules in each of which the protein moiety is different but the prosthetic group is essentially the same. Here again we see the simplicity of constructing many functional molecules using a few kinds of basic components. Metalo-porphyrin prosthetic groups occur in hemoglobin, chlorophyll, the cytochromes of electron transport, enzymes such as catalase, and many others. In every case there is a different protein portion in the molecule, owing to different informational specificities of the genes, but the prosthetic groups are fundamentally similar in nature. Building upon a common prosthetic group, large numbers of functionally distinct molecules could have been incorporated into the improving metabolic machinery of primeval populations. Each new protein would confer a new specificity on a molecule, such as an enzyme, thus expanding the repertory of cellular chemistry. We must assume that an increasing array of such molecules were inserted into closely coordinated and interacting groups of metabolic reactions. There would be little advantage to any single improvement if the modification disrupted other reaction series in metabolism. Each step in the evolution of cellular chemistry must be viewed from the standpoint of the unity and regulation of reaction sets, as well as its potential to enhance synthesis or breakdown of chemical constituents in primeval organisms.

Metabolic Diversity

Although there are a number of variations, three basic types of metabolism are recognized in life forms, depending on the sources of energy and carbon atoms used for growth, repair, and reproduction. *Heterotrophs* utilize organic compounds for energy and carbon; *chemotrophs* derive energy from inorganic matter but require organic chemicals for their carbon source; and *autotrophs* derive both energy and carbon from inorganic materials. Familiar examples of heterotrophs are found throughout the animal, fungal, and bacterial groups of organisms. In each case the species secures necessary energy from reactions involving degradation and synthesis of organic chemicals and obtains its required carbon for building new molecules from these same activities. Organisms differ relative to the particular chemicals that can be utilized, because each reaction is mediated by specific enzymes produced by the action of specific genes. But there are numerous organic chemicals, and each organism is capable of deriving some value from a sufficient variety to provide for life needs. Autotrophs occur principally among plants and such simple life forms as algae; the green leaf cells are those with which we are most familiar. Energy is obtained from specific wavelengths of solar radiation, and the carbon dioxide molecules of the atmosphere supply the principal carbon source for autotrophic life. Some kinds of bacteria also are autotrophic, but these represent only a small percentage of the multitudes of bacterial species now in existence. Chemotrophs are relatively rare forms of life today and are found mainly among groups of bacteria. The bacteria that derive energy from the oxidation of iron or sulfur compounds, among others, have been studied more extensively. While their energy is secured from inorganic compounds, carbon is derived from such organic chemicals as alcohols, acids, and various simple organic wastes released by other organisms. Variations on these three basic themes include life forms that are identified according to the particular energy source exploited; for example, chemoautotrophs and photoautotrophs both utilize inorganic energy and carbon sources, but the former group relies on chemicals whereas the latter absorbs light energy.

In addition to categorizing organisms according to their mode of nutrition, we also recognize differences in the ability of species to carry on metabolism in the presence or absence of molecular oxygen. The principal cell types in existence today are *aerobic,* requiring and utilizing oxygen in metabolism. There are some species of bacteria that can live only in the absence of oxygen; that is, they are obligately *anaerobic.* Most organisms are neither obligately aerobic nor anaerobic, however, and have the ability to perform some part of their metabolism under one set of conditions or the other. Such types are facultatively aerobic or anaerobic, depending upon which of these alternatives is the predominant pattern displayed.

Given these variations in present-day life, we must ask about the metabolic identification of the primeval organism. Once we can establish that point we can proceed to explore the evolutionary pathways and time periods that led to the spectrum of modern cell types. It is widely believed that primeval life was represented by *anaerobic heterotrophs,* that is, by forms that derived energy and carbon from organic sources and accomplished these activities in an atmosphere lacking molecular oxygen. There are several cogent lines of argument and evidence in support of this belief that the earliest life forms were a type that is relatively rare today. To take these points in some logical order, we must see at least one reason for considering the early organisms to have been anaerobic. Very simply, we believe that the primeval atmosphere was reducing in character and lacking in significant amounts of molecular oxygen. If we accept this premise then we must accept the corollary that the only life that could exist under these conditions would have been anaerobes. Whether they were obligately or facultatively anaerobic is a difficult detail to analyze and perhaps of no particular importance at this stage of the discussion.

Another set of arguments in favor of the proposed primeval anaerobic heterotrophs is of a more general nature. In fact, the basis for interpretation and decision is sufficiently general for us to use the arguments to discuss many features of primeval metabolism, including mode of nutrition. We accept the basic tenet of evolution to be descent with modification for any traits, including those of metabolism. Thus we would expect to discern relationships and sequences of events from a comparative study of existing life. This approach is accepted in solving evolutionary questions concerning ancestral and descendant anatomical features, as in conventional studies of comparative anatomy and embryology, and the principles apply equally well to comparative biochemistry and physiology if we have sufficient evidence for interpretations. For metabolism we would establish the following criteria to locate a system in an evolutionary sequence: (1) A primeval system would be one that was present in all or most living forms, because its complexity and gradual incorporation into coordinated reaction series would have occurred in ancestral lines and have been modified by evolutionary processes during later evolutionary times, each step being incorporated only in the context of the entire system; and (2) processes that evolved later in evolution probably would be found only in some kinds of organisms and not in any others, and such processes might be displaced or superseded by an older or more generally used mechanism. Although we do find similar systems in widely different organisms and can trace their evolutionary development to show that each arose independently in time and in different lines of descent, such cases tend to be exceptions rather than the rule. It is very unlikely that chance events consistently led to the evolution of very similar or identical traits during evolution and much more probable that similarity betokens relationship and common descent. Those traits found in all life forms, therefore, would be the most ancient, since they would be traced back to the earliest common ancestors. In the case of metabolism, these common ancestors may well have been among the earliest forms of life on the Earth. Similarly, if we detect alternatives in organismal metabolism we can be reasonably sure that one of these alternatives arose later in evolution. We simply are saying that any alternative that occurs in some kinds of life and not in others probably appeared later in evolution, since it was transmitted only to some descendant lines and not to all life. Any trait that arose earlier would have been transmitted to all or most later life forms, since these all have a common descent that can be traced back to more ancient times.

Using these criteria we can search for metabolic characteristics that occur universally and can be considered more ancient than others that appear only in some cell types. Because we find some level of anaerobic metabolism in all cells, or at least find some of the enzymes associated with anaerobic metabolism, we may use this information as further support for our hypothesis that the primeval life forms were anaerobes. Although there are some forms of bacteria that exist under strictly anaerobic conditions, even this observation strengthens the premise, since we do not find the converse, that is, strictly aerobic cells lacking all traces of an anaerobic metabolic capacity.

One universally occurring system is that of glycolysis in which carbohydrates are broken down to simpler molecular endproducts. In view of the metabolic capacity or of the presence of glycolytic enzymes in all organisms, it is probable that this heterotrophic aspect of metabolism is primeval in origin. Another heterotrophic capacity of all cells is the participation of coenzyme A in reactions that lead to linkage between carbon-containing fragments to produce longer carbon chains. The carbon–carbon condensations are mediated by a variety of coenzyme A–activated compounds, especially of acetyl coenzyme A. Acetate is a simple 2-carbon molecule that usually is chemically inert, but it is activated upon conversion to an acetyl group by coenzyme A under enzymatic control. Acetyl coenzyme A is a connecting link in the metabolism of proteins, lipids, and carbohydrates in all cells, and probably is primeval or very ancient in origin. In contrast with universally occurring heterotrophic reaction systems, chemotrophic and autotrophic metabolism occur only in some kinds of cells and not in others. Furthermore, organisms that are primarily autotrophic or chemotrophic also contain heterotrophic abilities. For example, many green cells can be grown under conditions in which all the energy and carbon for metabolism are derived from organic compounds. The contrary does not occur; that is, heterotrophic cells cannot live autotrophically under any conditions.

GENETIC THEMES IN METABOLIC EVOLUTION

The primeval anaerobic heterotrophs utilized the organic compounds that had accumulated on the early Earth in the times before life appeared. There must have been relatively few genes in these organisms, and so very few proteins would have been manufactured according to specific genetic instructions. As long as the supply of organic chemicals lasted there would have been little particular advantage to mutants capable of utilizing a wider variety of raw materials for their life activities. From experimental studies of abiotic syntheses it seems clear that substantial amounts and kinds of ready-made organic compounds, including ATP, would have been present in the primeval seas. These chemicals could have been metabolized in a few simple steps and incorporated into cellular structure and function with little or no modification. But as life multiplied there must have been crises in various populations because of the dwindling supplies of necessary molecules. In populations lacking sufficient amounts of raw materials in their environments there would have been changing selection pressures relative to the changing requirements of such populations of organisms. At such times there would be advantages to any spontaneous mutants able to manufacture some of their requirements from simpler chemicals in the surrounding medium. Such simpler chemicals could not be utilized by the nonmutant members of the population nor could they be modified to some usable form. Of course, any mutants would have retained the original metabolic abilities of assimilating a complex chemical, but would have amplified its capacities by the acquisition of a pathway that produced this complex chemical from a simpler precursor molecule. Such evolutionary improvements may have occurred only in some populations, so genetic diversity would have been increased among primeval forms. If essential raw materials occurred in ever-decreasing supply in some populations that happened to produce no mutants with enhanced metabolic capacities, then extinction probably would have been their eventual fate.

A general hypothesis proposed by Norman Horowitz in 1945 provided an explanation for increasing metabolic complexity in evolving life forms, based upon biochemical genetic analyses. These basic studies related the synthesis of enzymes to the presence and activity of specific genes and have been important in enhancing our understanding of this relationship since the 1940s. Before the dawn of biochemical genetics there was the simplistic belief that genetically simple organisms were those that had simple nutritional requirements and that heterotrophic species were the most complex because of their requirement of many complex organic chemicals for existence. With this earlier notion it was assumed that autotrophic life was primeval and that heterotrophic forms were derived during subsequent evolutionary history. Autotrophs such as green plants require very little for their existence, just some sunlight, water, and the carbon dioxide of the air, along with various minerals. This life style was interpreted incorrectly to indicate that nutritional simplicity corresponded with genetic simplicity. With information from biochemical genetics it became clear that species that could manufacture all their complex molecular requirements from a few simple kinds of raw materials were species with an extensive enzyme repertory. The synthesis of many sorts of proteins, carbohydrates, lipids, and nucleic acids requires mediation of tremendous numbers of chemical reactions by enzymes that are specific for each of these reaction steps. On the other hand, the more complex the nutrient required for life, the fewer reactions necessary to complete the processing of the final product to be assimilated into the growing cell. Since enzyme proteins are gene products, a greater number of enzymes indicates a greater number of genes and, therefore, a more recent evolutionary history in general. This does not imply that all heterotrophs are very simple creatures by any means; far from it, in view of the fact that the highest forms of animal life are nutritionally heterotrophic. What the generalization did provide was the basis for understanding the evolution of metabolic variety from the standpoint of the relation between increased genetic information and increased numbers of protein products, especially of enzyme proteins specified by the genes.

As stated earlier, organisms in the primeval seas required various chemicals for growth and reproduction and had a handy supply of accumulated products of abiotic syntheses. As life multiplied and diversified, various kinds of molecules became depleted in one or more populations because abiotic syntheses do not proceed at a sufficiently rapid rate to replenish the materials assimilated by proliferating life forms. If members of a population required compound "G" for life and if compound "G" became depleted, the population would become extinct unless there happened to be mutant individuals with the genetic ability to use an immediate precursor, let us call it compound "F," to manufacture the needed "G." Organisms that could use either "G" or "F" (to make the "G") would survive in the depleted environment and continue to reproduce. Since they would leave descendants genetically like themselves, the population would change from principally types using "G" to ones able to use either "G" or "F." As long as "F" remained plentiful the population would continue its successful existence. The mutation that permitted the manufacture of "G" from "F" is assumed to have controlled the synthesis or activity of an enzyme, f, which catalyzed the reaction in which "G" would be produced in the cell. The gene specifying enzyme f would be transmitted to descendants, who also would remove "F" for their life needs. Clearly, compound "F" would diminish in amount, in the same way as "G" had done with the original population. As the environment changed again, there would be selective advantage for any organism that happened to contain the genetic potential to synthesize enzyme e, which could catalyze the formation of "F" from its precursor, "E." Ultimately, "G" would be produced in stepwise fashion, and the population would change to one with the greater enzyme repertory because such organisms could use any of several chemicals to make the essential endproduct. As life forms continued to increase in numbers and kinds, there would be an increasing premium for those populations that could manufacture essential molecules from ever simpler precursors, depending upon the availability of raw materials in the environment, of course. Each additional reaction step requires a particular enzyme specified by genetic information, and the acquisition of each new metabolic capacity occurs in a sequence that has been called "evolution backwards." The earliest primeval life contained fewer genes, fewer enzymes, and fewer metabolic abilities and therefore was more dependent upon the surroundings to furnish molecules either in finished forms or in forms requiring relatively little modification to yield the final endproduct. Metabolically more complex life forms thus are presumed to have arisen as a function of selectively favored gene changes that permitted less dependence on a limited external supply of metabolites. The evolution backwards scheme proposed by Horowitz can be summarized as follows:

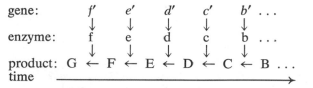

The proposal of evolution backwards should not be construed as meaning that metabolism can be improved only by an increasing ability to use a greater variety of organic compounds from the environment in which the population lives. Considering the metabolic pathways in modern cells, we can make the reasonable inference that genetic mutations in primeval life also permitted cells to couple reactions and reaction sets so that breakdown products of one metabolic pathway might be retained and routed to other internalized reactions. The products of such coupled reactions could be modified and serve as precursors for synthesis of some life requirement. In the same way that we may envision evolutionary development to permit increasing independence of the supply of chemicals in the external environment, another dimension of independence would be conferred on organisms that did not eliminate all reaction endproducts. Such organisms would have an internal supply of precursor molecules, which might be a more reliable and continuing source of raw materials than could be secured in some kinds of environments.

Modern cells have many sets of metabolic pathways in which the processes of synthesis and breakdown are coupled and coordinated in time and in space (Fig. 4.3). Not only are chemical products channeled from one reaction series to another with efficiency, but energy retention is more likely in such systems. Each incompletely processed molecule of a breakdown reaction contains unreleased energy. If this molecule is shunted to a reaction involved in synthesis, then that chemical energy is available for the process and represents a more efficient metabolic event. In addition, the coupling of reaction series permits the storage and eventual use of

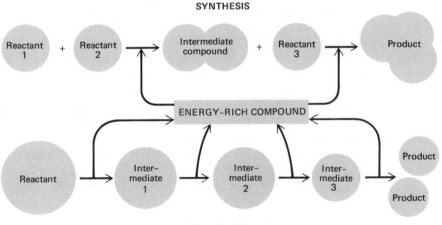

Figure 4.3 The synthesis and breakdown of reactant molecules usually are coordinated. Energy released in breakdown reactions may be stored in energy-rich compounds (such as ATP) and utilized in coupled synthesis reactions.

packets of energy that are released in one set of reactions. The release of energy in small packets not only allows for storage in chemical bonds but also prevents the explosive release of energy, which would occur if all of it were to be released in one or a very few steps during the processing of a metabolite (Fig. 4.4). It is the occurrence of coupled reaction series, co-ordinated sets of reactions, and the release of energy in small and usable packets during many steps between the beginning and the end of a metabolic pathway that permits life as we know it on Earth. The remarkable efficiency of chemical processing in living systems also is a function of the many different kinds of molecules that can be manufactured from a relatively small number of building blocks. As we saw earlier, the infinite variety of proteins is derived from combinations of about 20 or more kinds of amino acids, and a limitless number of genes can be derived from combinations of only 4 kinds of nucleotides. Such economy and flexibility of biochemistry is not found in the nonliving world.

The general theme of improvement in metabolic flexibility as described for increasing diversity among heterotrophs serves equally well in understanding the advantages to mutants that could utilize different sources of energy and carbon-containing chemicals. As the heterotrophs used up more and more of the organic compounds in the environment, there would have been selective advantages to life forms genetically capable of deriving energy from more plentiful and available sources. Mutants with chemo-

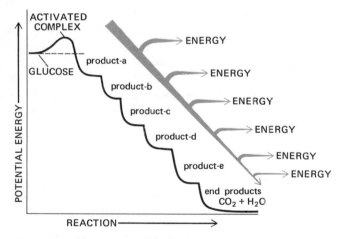

Figure 4.4   The energy contained in the glucose molecule is released in packets during the stepwise processing of the molecule in metabolism. The maximum energy release occurs when glucose is completely oxidized to molecules of carbon dioxide and water. Some of the energy is lost as heat, and the remainder is stored in energy-rich compounds, such as ATP.

trophic or autotrophic abilities, superimposed on the ancient heritage of heterotrophy, surely would flourish in any environment in which they could secure their life requirements. Many kinds of mineral compounds were present but used very little or not at all by existing heterotrophs, so that mutants with alternative nutritional modes would have been able to endure the depletions of energy-rich organic chemicals. Also, with increasing numbers of heterotrophs in the primeval seas there would have been accumulations of organic waste products of metabolism, most of which could not be utilized further by existing heterotrophic species. Chemotrophs not only derive energy from inorganic chemicals; many are able to obtain carbon for synthesis from such common metabolic wastes as alcohols and acids.

Of course, such evolutionary change would not occur in all environments or populations, because selection pressures would vary considerably, as they do on the Earth now. Many different kinds of habitats prevailed in ancient times; therefore many heterotrophic forms would have continued to flourish and evolve toward greater variety, just as new metabolic activities might have arisen and multiplied in other populations. The appearance of a new form in evolution usually does not lead to the total extinction of all earlier species. If that were the case, we would find no reptiles on the planet after mammals evolved, or no bacteria after the appearance of multicellular life forms. Clearly, life exists in great diversity because of genetic changes that have occurred throughout evolutionary history; but the many kinds of environments and habitats, which contribute to natural selection relative to the gene pool of populations, also contribute to its diversity.

As chemotrophs evolved with changing selection pressures in some environments, we would expect autotrophic species also to appear. Photoautotrophic types can utilize the energy from the abundant and continuous supply of solar radiation and, like chemoautotrophs, can build all their cellular constituents from inorganic carbon sources such as the carbon dioxide of the air. Each type requires relatively little from the surroundings, but each contains an elaborate enzymatic machinery to process these simple raw materials into the complex organic molecules that ensure growth, maintenance, and reproduction.

SOME IMPORTANT METABOLIC INVENTIONS

Although we are not completely certain, it seems most probable that the metabolism of the increasing numbers of chemotrophs and autotrophs would have led to an increase in the molecular oxygen content of the atmosphere. Relatively few enzymes are required for conversion from anaerobic to aerobic pathways in modern organisms, and relatively few initial changes would have been required for primeval life to switch to a diversity of life styles in which oxygen was involved. Many present-day chemoautotrophic species are found in habitats in a zone that is transitional between reducing and oxidizing conditions. As oxygen becomes a significant component of the atmosphere, reduced organic compounds from the underlying layers of the Earth become oxidized upon contact with gaseous oxygen. Organisms such as the sulfur, iron, and nitrifying bacteria almost always are found in those transitional regions in which oxygen occurs in relatively low concentrations. Their underlying heterotrophic capacities, upon which chemotrophy or autotrophy have been superimposed, indicate their position in the evolutionary ranking to be derived forms with a primeval heterotrophic ancestry.

Photoautotrophs probably evolved after the appearance of a considerable amount of metabolic diversity had become established on the Earth. For one thing, we know that some of the simplest autotrophs are principally anaerobic, whereas more highly evolved forms are generally aerobic types. Photoautotrophs possess extremely complex synthetic machinery by which a few simple kinds of molecules can be processed into all the varieties of organic compounds of the cell, these reactions being mediated by enzyme translations from genetic information, of course. Each added capacity in synthesis requires time during which mutations accumulate and selection pressures modulate the genetic complexion of a population. Furthermore, if we accept the assumption that the early atmosphere was reducing, then we must accept the later appearance of photoautotrophs, since these organisms are responsible for the oxygen levels in our present atmosphere. Once the seas became populated with a significant variety of photoautotrophic life, the atmosphere rapidly assumed its present composition. There must have been a considerable change in life forms, because many would not have survived the crisis of a change in the gaseous envelope of the planet. The preponderant life style of modern species is aerobic, and many or most present-day forms could never have evolved in an anaerobic environment. Life on the land, for example, would be entirely different if oxygen were absent. Some of the reasons will be discussed in the context of respiratory metabolism, but we will consider the evolution of photosynthesis first, because it probably preceded respiration in the history of life on Earth.

THE EVOLUTION OF PHOTOSYNTHETIC SYSTEMS   All known photosynthetic cells have at least one feature in common, namely, the production of ATP solely at the expense of light energy. This process is known as photophosphorylation; it is a cyclic set of events in every photosynthetic cell, but it also may occur in these same cells in a noncyclic variation. If we invoke the principle of universality as a criterion for recognizing more ancient origin, then cyclic photophosphorylation is the ancestral process and the noncyclic reaction series arose later in evolution. In the least advanced photosynthetic cell types photophosphorylation may be quite separate from processes of CO_2 fixation (the assimilation of CO_2 into carbohydrates) and the production of molecular oxygen. These events, therefore, also may be considered as later improvements and enhancements of the photosynthetic process.

The sequence of evolutionary events that has been proposed by Daniel Arnon and others begins with the evolution of the chlorophyll molecule under the conditions of a reducing atmosphere on the primeval Earth. The formation of a magnesium–porphyrin compound would be as probable an abiotic event as the formation of heme, which was discussed earlier. With the origin and evolution of life forms, mutation and selection could have led to the synthesis of the chlorophyll molecule in which protein is conjugated to the magnesium–porphyrin compound. Once chlorophyll molecules were manufactured in cells, such organisms became capable of capturing light energy and of converting this energy to a chemical form stored in ATP molecules which were products of cyclic photophosphorylation reactions. There would have been considerable selective advantage for such cell types because photophosphorylation is a more efficient mechanism for ATP production than the fermentative processes of anaerobic metabolism that occur in heterotrophic pathways. Fermentative reactions occur in modern cells, another vestige of our anaerobic past, but additional means of ATP synthesis are superimposed on the heterotrophic capabilities. The bacterium *Chromatium* is representative of primeval photoautotrophs, inasmuch as ATP synthesis in the light is independent of auxiliary photosynthetic processes.

Cyclic photophosphorylation is a self-contained process; the electrons released from the chlorophyll molecules under the influence of light energy ultimately return to these molecules. During the circuit traveled by the electrons, energy is removed at two different points and used to couple adenosine diphosphate (ADP) to inorganic phosphate (P_i) in the production of ATP. This phosphorylation of ADP to form ATP requires the participation of electron acceptors. Among these are the ancient protein ferredoxin and various cytochromes. Cytochromes are molecules composed of protein conjugated to a metalo-porphyrin complex, and these also are believed to be ancient compounds. Obviously, even though photophosphorylation represents an improved mechanism for ATP synthesis relative to fermentative reactions, there is no reason to expect the new reactions to occur in all cells, since the particular mutations that led to chlorophyll manufacture did not occur in all cell lineages. By the time the chlorophyll molecule must have originated, many different heterotrophic species must also have been present. Undoubtedly only one or a few of these species happened to undergo the particular sequence of mutations that permitted the development of a photophosphorylation mechanism during which ATP synthesis occurred. But in photophosphorylation systems there were further modifications and improvements, which led to the addition of noncyclic mechanisms to some ancestral lines that already possessed the cyclic format (Fig. 4.5).

In the more ancient noncyclic photophosphorylation system, such as that in *Chromatium,* ATP is synthesized in the light. As in the cyclic process, chlorophyll molecules absorb light energy and release electrons. But these electrons combine with molecules and ions such as hydrogen, nitrogen, or pyridine nucleotides, and reduced compounds are formed. Since the chlorophyll molecules do not retrieve their lost electrons in this mechanism, an electron donor is required to replenish these molecules. In species such as *Chromatium,* organic compounds such as succinate or thiosulfate provide electrons, which are transferred from cytochromes to chlorophyll. In this sequence, some of this energy, plus protons, is channeled to the manufacture of ATP from ADP plus inorganic phosphate radicals. The reduced pyridine nucleotides, another group of ubiquitous organic compounds, participate in many metabolic pathways, including the so-called dark reactions of photosynthesis in which CO_2 is fixed and then assimilated into carbohydrates. Molecular oxygen is not produced in this type of noncyclic sequence; it is a product of the noncyclic photophosphorylation pattern typical of the more highly evolved green plant cells.

In green plants there is a much more complex system of reaction processes, which involve both chlorophyll *a* and the more recently evolved chlorophyll *b* molecules. The synthesis of ATP in the light is a noncyclic event, but a significant modification has occurred in the nature of the donor of electrons and protons (see Fig. 4.5). Unlike the organic compounds utilized by *Chromatium* and similar organisms, water serves as the donor of electrons to chlorophyll and of protons for the reduction of pyridine nucleotides in green-cell noncyclic photophosphorylation. There are obvious advantages to the use of water as a donor. In the first place, water is a ubiquitous compound, so green cells are essentially unrestricted, whereas primitive forms such as *Chromatium* can manage successfully only in habitats containing a ready supply of organic chemicals such as thiosulfate. Since the primeval seas were depleted of many organic compounds, it is obvious that these molucules were a limiting factor to population success. A second advantage to the donation of electrons and protons of the hydrogen atoms from water is that the split of the water molecule in the light led to the release of molecular oxygen as a by-product of the reaction. Oxygen release became an inseparable part of green plant photosynthesis once their noncyclic photophosphorylation process included the photolysis of water molecules.

With the production of oxygen from water, the atmosphere of the Earth rapidly achieved an oxygen content of slightly over 20 percent, and it has remained at this level ever since. Although the release of oxygen was incidental to the evolution of green cell photosynthesis, the other events of the process make up a sequential evolution of sets of reactions all of which

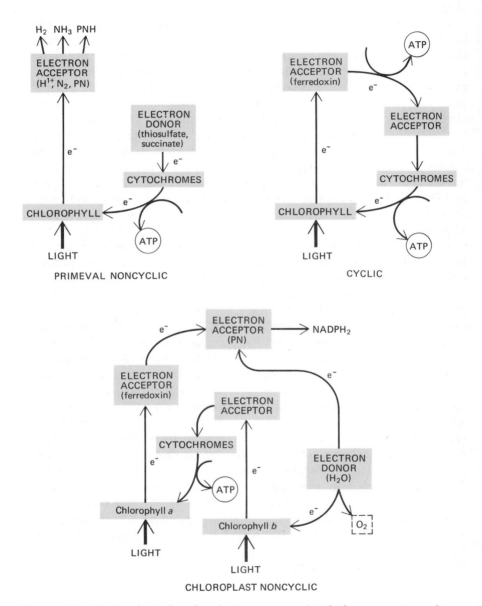

Figure 4.5 Cyclic photophosphorylation compared with the more primeval mode of noncyclic events, as in the bacterium *Chromatium*, and the more advanced noncyclic photophosphorylation reactions of green plant chloroplasts. The utilization of water as an electron donor leads to the release of molecular oxygen as a byproduct of the reaction in chloroplast noncyclic photophosphorylation.

occur together in modern species. These include the reactions of ATP production in the light, reduction of pyridine nucleotides in the presence of protons from water molecules, photolysis of water which provides these protons as well as the electrons that replenish the chlorophyll energy level, and the reactions of CO_2 assimilation which are not light-requiring.

THE EVOLUTION OF AEROBIC RESPIRATION   Once oxygen leveled off at about 20 percent in the atmosphere, there would have been strong selection pressures favoring mutants that could derive energy and also process degradation products of glucose in ways that were more efficient than by fermentation alone. Two new systems were required during the development of aerobic respiration: (1) a system by which hydrogen ions (protons) would be mobilized instead of being lost as waste products in such incompletely oxidized fermentation endproducts as acids and alcohols, and (2) a system for activating oxygen as the final acceptor of protons, yielding water as the innocuous endproduct of the aerobic respiratory sequence. Coupled with these major systems are reactions that yield ATP in much greater abundance than occurs in fermentation alone. Indeed, aerobic respiration is the principal source of the ATP that is generated in all cells with an aerobic capacity, and life as we know it would be impossible without this energy-storing molecule.

Relatively few changes were necessary to modify existing anaerobic systems, because hydrogen acceptors such as the pyridine nucleotides and coenzyme A already were present. An example of an improvement in the hydrogen transport system occurs in facultatively anaerobic bacteria as compared with obligately anaerobic species. In anaerobes the key molecule pyruvate undergoes anaerobic dehydrogenation and decarboxylation to yield acetate, lactate, and ethanol as waste products. Facultative anaerobes can oxidatively decarboxylate pyruvate to produce acetate, which then undergoes condensation to yield oxaloacetate. Oxaloacetate is one of the reactants in the citric acid cycle (also known as the tricarboxylic acid cycle or the Krebs cycle), which is an integral part of aerobic respiration (Fig. 4.6). The inclusion of the citric acid cycle required the incorporation of steps that accomplish the transfer of protons via pyridine nucleotides and dehydrogenase enzyme activities.

The cycle picks up from glycolytic reactions of fermentation with the condensation of activated acetate (as acetyl coenzyme A) and oxaloacetate to form the 6-carbon molecule citric acid. Reactions proceed in cyclic fashion and oxaloacetate is regenerated, thus permitting further condensations with new molecules of acetyl coenzyme A to continue the cycle. The citric acid cycle allows for the utilization of all the energy of the hydrogen atom and provides a direct connection at several points with the metabolism of proteins, lipids, and carbohydrates. The stepwise transfer of hydro-

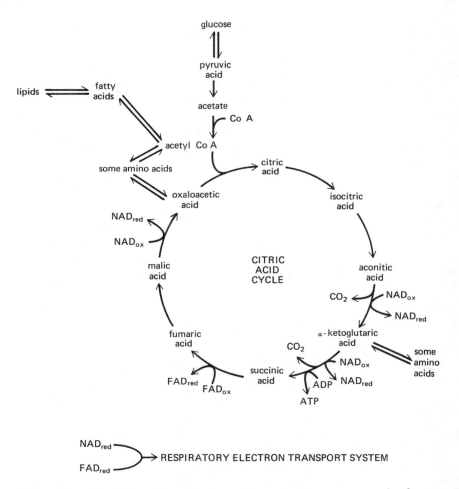

Figure 4.6 The citric acid cycle of aerobic respiration, a crossroads of metabolism.

gen overcomes the energy barrier and, at the same time, permits the release of energy in small, usable packets rather than in an explosive release. A great deal of this energy is packaged in ATP molecules during each turn of the citric acid cycle; the remainder can be released and packaged during subsequent reactions, such as the coordinated set of events leading to oxygen activation. The oxygen activation system, usually described as the electron transport chain of aerobic respiration, is coupled to oxidative phosphorylation reactions in which ATP synthesis occurs (Fig. 4.7). We will consider this system next.

The combination of gaseous hydrogen and oxygen does not occur at cell

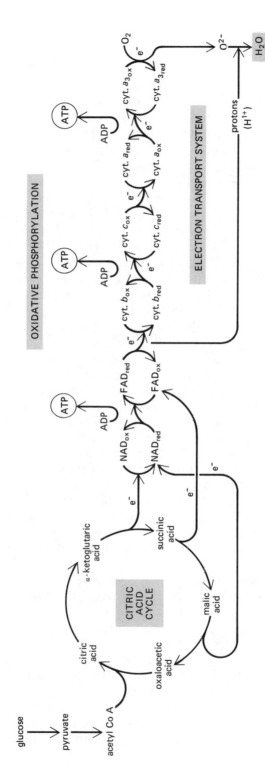

Figure 4.7 Hydrogen atoms released during each turn of the citric acid cycle are processed such that their electrons are transported along a respiratory chain of nicotinamide and flavin adenine dinucleotide (NAD and FAD) coenzymes and a series of cytochromes which activate molecular oxygen to accept protons released earlier, yielding molecules of water as endproducts. The synthesis of ATP in the coupled reactions of oxidative phosphorylation also are shown.

temperatures because of the very high energy of activation of this reaction. The combination of hydrogen and oxygen atoms can occur if there is a stepwise transfer of hydrogen and a gradual release of the contained energy. The link between the hydrogen transport system of the citric acid cycle and the transfer of electrons until the final combination with the atoms of molecular oxygen resides in the pyridine and flavin adenine dinucleotide enzyme and coenzyme systems. Through these systems the hydrogen is transferred to a series of oxidase and peroxidase enzymes of the respiratory chain. These terminal respiratory enzymes activate molecular oxygen and peroxides to accept protons, with the final production of water at the conclusion of the reaction series. The coupled ATP-generating steps of oxidative phosphorylation can be uncoupled from the oxygen activation system without impairing oxygen consumption. There is considerable variety in the oxidative reaction steps, leading to the general belief that some of the sequences are more recent in origin, having arisen independently in different species groups at different times in history. Cytochromes are the most widely distributed components of the respiratory chain and probably more ancient than other parts of the oxygen activation system.

Some obvious advantages accrue to organisms with an aerobic respiratory capacity. The respiratory system is a much more efficient mechanism of energy release and of utilization of organic compounds than is fermentation. Aerobic respiration permits the organism to retrieve much more energy and many more building materials for growth and proliferation than the slower and less efficient fermentative pathways. Also, water is released as an endproduct, rather than such toxic wastes as the acids and alcohols of fermentation. Casual observations of present-day life reveal that very few species are exclusively or principally anaerobic, but many anaerobic reactions characterize modern cells. Indeed it is doubtful that life could have evolved on the land as well as in the water unless efficient pathways had been available for production and storage of chemical energy, an essential component for the diverse forms of life and especially for large organisms.

THE OZONE LAYER   In addition to subsidizing respiration, the introduction of molecular oxygen into the Earth's atmosphere led also to the development of a thin layer of ozone (O_3) in the upper levels. Ozone forms from oxygen under the influence of ultraviolet radiation from the sun and from space. The blanket of ozone prevents the shorter wavelengths of ultraviolet radiation from penetrating through to the surface of the Earth. Since these wavelengths are the ones more harmful to life on the planet, being the principal germicidal and mutagenic wavelengths, life on the land could not have developed as it did had it not been for the ozone layer. Aquatic life is less subject to the hazards of shorter wavelengths of ultraviolet radiation because water absorbs the energy and dilutes the radiation that actually reaches aquatic life.

The thin layer of ozone in the upper atmosphere is maintained by a two-part mechanism: Ozone is converted to oxygen under the influence of shorter ultraviolet wavelengths while oxygen is converted to ozone in the presence of longer wavelengths, as follows:

$$O_2 \underset{\text{shorter wavelength UV}}{\overset{\text{longer wavelength UV}}{\rightleftharpoons}} O_3$$

It has been estimated that the atmosphere became oxidizing about 700 to 1000 million years ago. Thus, the Earth's atmosphere was reducing for the first 3.5 to 4 billion years of its existence.

Increase in Genes

If we believe that the early life forms were capable of limited protein synthesis and that their metabolic capacities and diversity increased with time, then we must assume also that there was a parallel increase in the numbers of genes and in the varieties of genetic information as life evolved toward greater complexity and differentiation. We know that mutations lead to modified genetic information, so new amino acid composition and sequence would result from alteration in the codons of the genes. Such mutational change may occur as additions and deletions of nucleotides and as substitutions of nucleotides. But it seems unlikely that additions of one or a few nucleotides would suffice to explain the increase in genetic information on a scale appropriate to the magnitudes of time and change that characterize biological evolution. It is clear that substantial changes leading to increased numbers of genes must have occurred, and considerable experimental evidence has been reported showing that this indeed could have taken place in the ancient past as well as in more recent times.

Although it is possible that the primeval life forms contained free genes, not aggregated end to end in multigenic linear arrays, it seems unlikely. Many studies of different sorts of linear macromolecules, including nucleic acids, have shown quite clearly that there is a strong tendency for end-to-end alignment to form long-chain polymers. If we examine the simplest genetic systems among the viruses we find that their genes always are contained within a single nucleic acid molecule and never occur disassociated. Indeed a single nucleic acid molecule comprised of all the genes of the species turns out to be the standard genetic pattern in every virus and bacterial species that has been studied, whether there are a few genes or hundreds or thousands. This consistency of gene grouping is a universal attribute in modern species with simple genetic systems, that is, among the known prokaryotes; therefore, it must have become established early in evolution and must have carried substantial selective advantage. We will discuss this aspect of the genetic system in the next section when we consider the evolution of more complex cell types from the simple bacterialike prokaryotic cell.

It would be reasonable to assume that primeval organisms had relatively few genes and that gene increases must have occurred. We also would predict that individual genes were shorter lengths of information in ancient life forms than in modern species. There have been very few studies directed toward the specific question of increasing lengths of individual genes, but the evidence that has been reported is interesting and has a direct bearing on the problem of increase in genetic information since primeval times. It is not feasible, generally, to examine the gene directly by analyzing nucleotide composition and sequence, but it is possible to study the amino acid content of proteins. Since the protein is a co-linear product of gene action, we may infer that changes in the protein reflect changes in the genetic information that specifies that protein. The protein of cytochrome c has been studied in detail in about 40 different species; the differences in amino acid composition of this respiratory chain component are correlated with the evolutionary relationships among these species. Thus, among the 104 amino acids of the protein there are 40 differences between the molecules from yeast and from the horse, but only 12 amino acid differences among the 104 when we compare human and horse cytochrome c. This general correlation has been found consistently for other proteins studied in detail, and this finding provides strong support for the premise that there has been descent with modification; more closely related species share more inheritance than do those more distantly related.

With reference to the problem of increase in gene length (informational content), as well as change in information as evidenced in cytochrome c comparisons, among others, there is an intriguing case history recorded in the amino acids of the ancient protein ferredoxin. This molecule probably was a significant component of early metabolism by virtue of its stability, highly reducing nature, and participation in a variety of nonphotosynthetic reactions as well as in photochemical events leading to the channeling of energy to storage systems within the cell. There are only 55 amino acids in this protein, which is considerably less than the 135 to 150 amino acids considered to be an average number of many proteins. A detailed examination of the sequence of amino acids in ferredoxin indicates that there is a repeat segment to the molecule such that amino acids 1 through 29 occur in a very similar pattern in the remainder of the molecule. Such identity of amino acid type and sequence is not considered a likely event on the basis of random occurrence, as can be shown by statistical analysis. The more likely explanation is that the original gene was considerably shorter and that a duplication of much of this information arose. The original and the duplicated segments aggregated end to end to produce a new stretch of information specifying a longer protein molecule. If this is a valid interpretation of the modern ferredoxin, then it can serve as a model to explain the origin of additional genes in a complement (by duplications of existing nucleic acid segments) as well as of genes that contained more information as new codons aligned with existing ones to constitute a larger individual stretch of information. As mutations occurred with time, we would expect that different changes would take place in codons and groups of codons so that different parts of the new gene would become increasingly divergent. But for some proteins enough of the original amino acid composition has been retained that an interpretation of the past history of the gene specifying the protein product is feasible. There are some slight variations in the numbers of amino acids in cytochrome *c,* but the essential alterations occurred in a molecule of the same length in all species studied. Some 35 amino acid sites have not changed at all, such as the glycine residues at positions 1, 6, 29, 34, 41, 45, 77, and 84, and the tyrosine at positions 48, 67, 74, and 97. Other amino acid residues have undergone extensive substitution, such as those at positions 44, 58, 60, and 89. The three-dimensional structure of this molecule, together with the amino acid sites which bind to the heme group of cytochrome *c,* provide an excellent foundation for understanding the variable and invariant aspects of the respiratory chain protein.

Ferredoxin is an example of a protein that differs in amino acid composition in various species, as was true for cytochrome *c,* but there are hints of an ancient duplication of genetic information with subsequent mutational modifications in the originally identical duplicate regions. Another case in which gene duplication has been implied involves the immunoglobin proteins of vertebrates, which occur as light-chain molecules with a molecular weight of about 20,000 daltons and a heavy-chain form of 50,000 to 70,000 daltons. Analyses of the proteins and genetic inheritance patterns have suggested that the heavier molecule is a double- or triple-length version of the smaller immunoglobin. Here, too, the original duplicated information has become modified by subsequent mutational steps such that the sequences no longer are identical. In those instances in which evolution has not proceeded to the point of functional differentiation of ancestral and derived proteins and genes it is possible to explore the relationships and to seek those mechanisms that led to the observed variations. If we follow this line of reasoning still further, we should expect that many of the genes in present-day life have come to be so different from their original versions that they now specify quite different proteins. Although we cannot be sure of the origins of very different modern proteins, we can infer that similar events produced molecules of very different functions, since the patterns are distinguishable from the magnitudes of the specific modifications.

The few genetic studies that have been possible have provided evidence for the occurrence of genes in different locations in the chromosome complement which specify the same functional enzyme protein. Because we know that the linear information sequence in DNA coincides with the linear translational sequence of amino acids in the protein, we can assume that such duplicate genes have not yet undergone significant mutational changes; therefore, each gene locus still guides the production of the same functional endproduct molecule. The alternative explanation, that identical genes arose by chance in the evolution of these species, is so remote a possibility that we can reject it from our consideration. The statistical probabilities for identical genes to have appeared by chance are so low as to be improbable. Hundreds or thousands of identical events would have been required to yield two genes with sufficient codon similarities, in both composition and sequence of nucleotides, that their translations would be identical or nearly so. Genetic similarity implies relationship, at the level of the single gene as well as at the level of the entire genetic complement of species.

Recently, methods have been developed that allow us to make direct quantitative analyses of genetic repetition, a duplication phenomenon of much greater magnitude than those we have considered till now. From such studies we have discovered that repeated sequences of DNA are typical of all the eukaryotic species so far examined. Hundreds or thousands of copies of DNA regions of chromosomes occur, for reasons and advantages not yet apparent. But for such a universal phenomenon to be present in eukaryotes, it must have selective advantages. Future studies may shed light on this recently discovered aspect of genetic duplication in evolution.

Some Suggested Readings

Arnon, D. I., "The role of light in photosynthesis." *Scientific American* 203 (November, 1960), 105.

Blum, H. F., "On the origin and evolution of living machines." *American Scientist* 49 (1961), 474.

Britten, R. J., and D. E. Kohne, "Repeated sequences in DNA." *Science* 161 (1968), 529.

Dickerson, R. E., "The structure and history of an ancient protein." *Scientific American* 226 (April, 1972) 58.

Eck, R. V., and M. O. Dayhoff, "Evolution of the structure of ferredoxin based on living relics of amino acid sequences." *Science* 152 (1966), 363.

Oparin, A. I., *The Origin of Life on the Earth*. 3rd ed. New York, Academic Press, 1957.

Zuckerkandl, E., "The evolution of hemoglobin." *Scientific American* 212 (May, 1965), 110.

CHAPTER 5

PROKARYOTES AND EUKARYOTES

Prokaryotic organisms are those that lack an organized nucleus and have no nuclear membrane separating the naked DNA molecules from the general cytoplasmic materials of the cell. Included among the prokaryotes are the bacteria, mycoplasma, rickettsias, actinomycetes, and the blue-green algae. Eukaryotes have a nuclear membrane surrounding two or more chromosomes; thus the nucleus essentially is compartmented within the cell and is the site of the principal genetic material of the organism. Both kinds of life forms may exist as unicellular or as multicellular organisms, but prokaryotes do not attain the size or bulk of eukaryotes even when composed of many cells per individual. Except for the viruses, which are acellular and therefore neither prokaryotic nor eukaryotic, all other known forms of life can be included in one or the other of these two major categories. This fundamental distinction between two principal groups of life has been recognized as a significant divergence only in recent years. For many biologists the separation of organisms as prokaryotes and eukaryotes has much greater meaning in an evolutionary context than the emphasis on differences between plants and animals, which forms the basis for older and more conventional schemes of classification. Similarly, the older views emphasized distinctions between the heterogeneous collection of microorganisms, whether prokaryotic or eukaryotic, and the more complex forms of plants and animals. These earlier, simplified views of species groups were based more upon gross appearances than upon comparative biochemistry, genetics, and the general format or organization of cell contents and structures. This is not to say that we are less interested in the more familiar kinds of life; rather, we have a new perspective on evolutionary pathways, one that permits us to explore and to perceive relationships in ways that are somewhat different from the ways possible in earlier years.

In addition to the fundamental differences in the packaging of genetic material within a membrane-bound nucleus or in a nonmembranous nucleoid region of the cell, various other differences can be cited for these two forms of life (Table 5.1). In this chapter we will be concerned only with some of these characteristics, looking especially at relative advantages and evolutionary derivations wherever possible.

TABLE 5.1   SOME OF THE MAJOR DIFFERENCES BETWEEN PROKARYOTIC
AND EUKARYOTIC CELLS AND ORGANISMS

Characteristic	Prokaryotic	Eukaryotic
Genetic system	Duplex DNA not associated with proteins in chromosomes	Duplex DNA complexed with proteins in chromosomes
	Nucleoid not surrounded by a membrane	Membrane-bound nucleus
	One linkage group	Two or more linkage groups
	Little or no redundant DNA	Redundant DNA
Motility organelle	Simple flagella, if present	Complex flagella or cilia, if present
Internal membrane system	Transient, if present	Numerous types: endoplasmic reticulum, mitochondria, chloroplasts, lysosomes, peroxisomes, etc.
Tissue formation	Absent	Present
Cell division process	Fission (or other means)	Mitosis
Sexual system	Unidirectional transfer of genes from donor to recipient	Complete nuclear fusion of gamete genomes (equal contributions)
Nutrition	Principally absorption (some photosynthesizers)	Absorption, ingestion, photosynthesis
Cell size	Generally smaller (1–10 μm)	Generally larger (10–100 μm)

The Prokaryotic Genetic System

All known species, including the smallest viruses, contain genes in structural aggregates. Genes contained within a common linear grouping are considered to be *linked* to each other, and such an assemblage is a genetically detectable *linkage group*. From genetic analysis it is possible to determine whether genes under study are inherited independently from one another or whether such genes do not undergo independent assortment. Genes that show independent assortment, as was explained by Gregor Mendel more than 100 years ago on the basis of statistical arguments, are on separate structures; those that tend more often to be inherited together than independently are found to occur in the same structure or linkage group. Inasmuch as no system is known to have genes that are entirely unassociated—that is, free genes rather than linked genes in one or more structures—there must be considerable selective advantage to the organization of genes into a common physical entity. The unique feature of the prokaryote genetic arrangement is that all the genes of the species are contained within a single naked DNA molecule which is the cytological equivalent of the genetic linkage group. The length of this one DNA molecule may vary from a few micrometers in the smallest viruses to more than

1000 micrometers (one millimeter) in a common bacterial species such as the intestinal bacillus *Escherichia coli* (Fig. 5.1). Genetic studies have shown that the length of the DNA molecule is directly proportional to the number of genes in the species and that relatively little, if any, nongenetic material exists in this molecule. Since we know that all the genes are part of one DNA molecule in the prokaryotes that have been studied, there must be strong selective advantages not only to a linear aggregate of genes but also to the existence of just one linkage group and no more.

There is at least one observation that provides a basis for explaining the advantages of end-to-end aggregations of genes and the concomitant phenomenon of a single DNA molecule containing all the encoded genetic information. Prokaryotes lack a mechanism for ensuring the precise and equal distribution of genetic material to the daughter cells during fission. There is no mitotic mechanism present as in eukaryotes. According to current information, the separation of newly duplicated DNA molecules occurs in more or less passive fashion. The DNA is attached to the cell membrane; as this membrane of the dividing cell lengthens, the daughter molecules are drawn farther and farther apart from each other. When the wall that separates the daughter cells forms, each new cell contains a copy of the genetic information, since the DNA molecules are well separated from each other by that time. It is a reasonable inference that such a mechanism would be most successful if all genes were part of one molecule rather than existing free from each other or in more than one piece of DNA. There is less chance for loss of genes, and this is important in cells containing little or

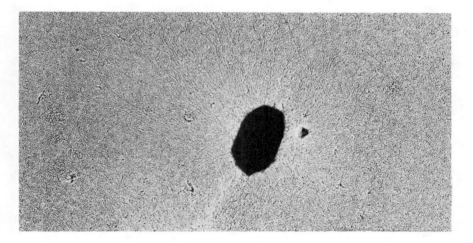

Figure 5.1   A cell of the bacterium *Pseudomonas* was exposed to osmotic shock by spreading on a water surface. The burst cell has released its naked duplex DNA. (Electron photomicrograph courtesy of H.-P. Hoffmann.)

no redundant information. Each gene is necessary for the successful growth and reproduction of the cell, and there would be strong selection pressure to preserve any mechanism that allowed the transfer of a complete gene set during each fission event. Any mutant that failed to transfer all its genes faithfully would be at a severe disadvantage and would soon be lost from the population. Types with the genetic properties of maintaining all genes in a single molecule and of transferring this molecule intact in each generation would be successful and would transmit these inherent successful traits to offspring down through the generations. It is also clear that new genes could be added to an existing linkage group and be inherited, as part of the single DNA molecule, by each succeeding generation.

Gene recombination occurs in prokaryotes and leads to greater genetic diversity in populations than can occur by mutations alone. Little was known about recombination processes in prokaryotes until recently, but genetic studies since 1946 have provided abundant evidence that genes from different cell lineages come together and provide the possibilities for genotype recombinations to occur. The molecular mechanism leading to recombination involves a process of breakage and reunion of homologous DNA segments, and it may be quite similar in all life forms, although the evidence is strongest for such a process at least in prokaryotes and viruses. The crucial requirement, however, is to bring together the genes from different lineages so that the opportunity is present for the recombination events. Microbial genetic studies have demonstrated the existence of various mechanisms by which genes from one cell line may be introduced into another cell line and lead to the production of recombinant progenies. By phenomena such as transformation, transduction, and conjugation some genes from one lineage are introduced into a receptor cell. There are at least two significant features of gene transfer mechanisms in prokaryotes: (1) Only some of the genes of the donor are introduced into the recipient cell—rarely if ever are all the donor genes transferred; and (2) there is no regularity for this unidirectional transfer of genes from donor to recipient cell line. In eukaryotes there is the fusion of gamete cells leading to a zygote nucleus containing all the genes of the donor as well as the recipient. In addition, in sexually reproducing species fertilization provides opportunities for gene recombination in each generation every time the gametes fuse, and for all the gametes engaged in such unions. In the prokaryotes, on the other hand, recombination may be a rare or a frequent event depending on the external conditions and controls of the population. The irregularity of opportunities for recombination probably yields a lower level of genetic diversity in prokaryotes, but since recombination is possible it does provide an important mechanism for adding genetic diversity to populations. This has helped us to understand better the reasons for prokaryote success during the billions of years of their evolutionary history, a success that would seem unlikely if mutations were the sole basis for inherent variation.

The Eukaryotic
Genetic System

One striking difference between eukaryotes and prokaryotes is that no known eukaryote contains fewer than two chromosomes as its basic complement, whereas the known prokaryotes have only one linkage group. In addition, the simple duplex DNA molecule of prokaryotes is not found in the eukaryote nucleus. Instead, eukaryote genes are contained within the DNA of the chromosomes, but either this DNA is so arranged as to appear a structural equivalent of a conventional chromosome or proteins and other chemical constituents are incorporated into the chromosome proper. Among most eukaryotes DNA is conjugated to proteins of the histone group; the basic structural plan involves nucleoprotein strands rather than either DNA or protein alone. Although some simple protists, such as the dinoflagellates, have no detectable protein in their chromosomes, the arrangement of the DNA fibrils results in a recognizable three-dimensional chromosome structure (Fig. 5.2). The conventional eukaryote chromosome is a complex differentiated structure with regions that are morphologically and functionally different from other parts of the same chromosome (Fig. 5.3). There also is a differentiation between chromosomes in the same nuclear complement, since each genome contains at least one chromosome with a nucleolar-organizing region.

There is no particular correlation between chromosome number and evolutionary rank of eukaryote species, nor does the amount of DNA per nucleus have any absolute bearing on distinguishing species groups distributed along evolutionary pathways. In contrast with these phenomena, the amount of DNA in viruses and known prokaryotes is directly related to the complexity and number of genetic units in the single DNA molecule of the species. Several reasons contribute toward an explanation for the lack of correspondence between number of chromosomes and evolutionary ranking of eukaryote species. Recent studies have indicated the occurrence of a substantial number of repeated sequences of DNA in the chromosome complement. Although we do not yet understand the significance of the observed informational redundancy, there is no way available at present to compare DNA amounts in eukaryotes, simply because there is so much repetitious DNA and only a handful of species have been analyzed in any detail for this quality. But as usual when such a universal phenomenon is discovered, we would expect to find a substantial selective advantage underlying the invariant occurrence of the characteristic in a major life form. Another reason for the lack of correspondence in eukaryotes is that the number of chromosomes is related not merely to gene numbers or complexity of the genetic program, but principally to the advantages of a particular genetic system to the whole life style of the species. In a later chapter we will discuss in some detail the features of the genetic system, especially in relation to chromosome numbers, which contribute to species

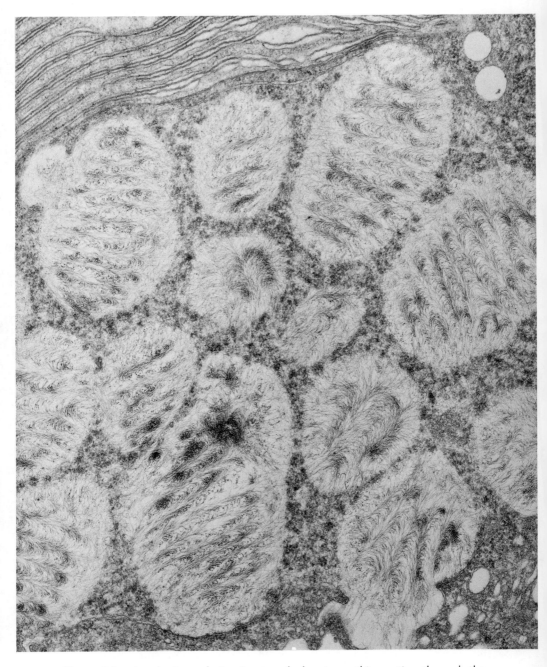

Figure 5.2 An electron photomicrograph showing a thin section through the nucleus of the dinoflagellate *Zooxanthelle*. The chromosomes consist of tightly packed naked DNA molecules, and are enclosed within a membrane-bounded nucleus typical of eukaryotic cells. (Photograph courtesy of H. Ris.)

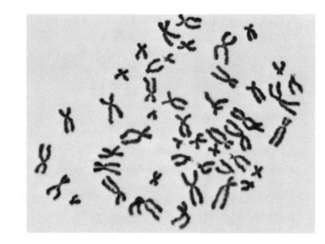

Figure 5.3 The human chromosome complement contains 23 pairs of chromosomes, which are differentiated morphologically. Each chromosome has a constricted kinetochore region, which is required for movement of the chromosome to the poles of the cell during mitosis and meiosis. (Photograph courtesy of L. J. Sciorra.)

success in the short run as well as from a long-term point of view. For now, it is pertinent to mention at least that chromosome number bears directly upon the recombination potential of a species.

In general, higher chromosome numbers permit greater genotype diversity, the level of diversity that is generated serves as one crucial criterion for evaluating the evolutionary potential of a species group. Thus, among animal species one may find almost any number of chromosomes in any of the different phyla, with many having fewer than 23 pairs found in humans, and other groups having many more than 23 pairs. None of these numbers is relevant to ranking the human species as more or less evolved than other animals. Indeed, chromosome numbers among the primates themselves bear little relation to evolutionary success or lack of success of these species. Some of the lowest chromosome numbers recorded belong to plants of the sunflower family, a group that is considered to be among the most highly evolved of all angiosperms (flowering plants). But many species in this same family have high chromosome numbers, so once again there is no correlation to be deduced for this particular eukaryote trait. Of course, as we shall see later, there are trends to be noted among closely related species in which change in chromosome numbers can be related specifically to level of species success during recent evolutionary history. But no generalizations can be stated for chromosome number and the progression of forms in eukaryote evolution.

In the same way that we searched for a principle to explain the universal occurrence of a single linkage group in viruses and prokaryotes, we should expect to find a basis for understanding the presence of chromosomes in any number from two to many hundreds in the eukaryote nucleus. If the problem in prokaryotes was based on the mechanics of chromosome distribution to progeny cells, then this problem must have been solved in the eukaryotes, since they all contain more than one chromosome. Furthermore, the solution to the problem of chromosome distribution must be such that the absolute number of chromosomes is not a limiting factor in the success of the distribution mechanism.

The obvious answer is the evolution of a mitotic mechanism by which replicated chromosomes, in any number, are distributed precisely to daughter cells at every cell division in every generation of cells (Fig. 5.4). Recent studies of a dinoflagellate species have revealed a very interesting modification of a chromosome distribution mechanism, and more variation on the theme of classical mitosis undoubtedly exists. In conventional mitosis there is a spindle apparatus comprised of microtubules, some of which are attached to the kinetochore region of the chromosome. Chromosome movement depends on the formation of these microtubules as well as on the presence of a kinetochore region; the lack of either leads to the failure of the chromosomes to separate accurately or to separate at all in some cases. Such a precise mechanism, which depends on synthesis and polymerization of microtubular protein and on the structural and functional differentiation of every chromosome as well as on many other factors of a coordinated cell cycle, could not have arisen at one step in evolution. Furthermore, with so many points of difference between prokaryote and eukaryote chromosome distribution mechanics, it would be reasonable to expect that a number of intermediate stages appeared during the evolutionary development of the modern system. It would be easy to say that any intermediate stages would have become extinct as further improvements were incorporated into the mitotic mechanism, and that prokaryote survival and success were func-

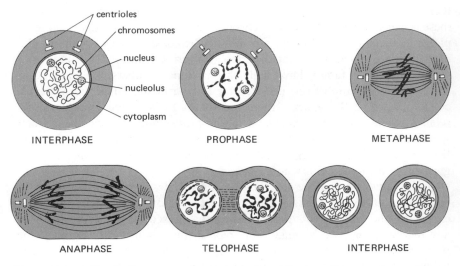

Figure 5.4   Mitosis. Sequence of events during mitosis as it occurs in most eukaryotes.

tions of their lacking altogether a variation on their own particular genetic system. And we did tend to think this way because no intermediate or variant forms were known.

With the report in 1969 of the dinoflagellate mechanism of chromosome distribution we became aware of the possibility that significant variations may exist and that additional observations might turn up just those intermediate types from which an evolutionary sequence might be reconstructed. Interestingly, the dinoflagellate chromosome is a variant, just as the distribution mechanism differs from classical mitosis. Their chromosomes are morphologically similar to conventional eukaryote chromosomes and exist in numbers greater than one (see Fig. 5.2). But the chromosomes are comprised of DNA which is not complexed to proteins, and they do not undergo the cyclic contraction and extension we know for the usual chromosomes of other kinds of eukaryotes. During nuclear division channels form in the cytoplasm and pass through the nucleus at various places. The rigidity of these channels is maintained as a function of microtubules, but these protein fibers do not penetrate into the nucleus at all. Chromosomes attach to the membrane at the periphery of the nucleus and appear to be transported to the poles of the cell by some mechanism that involves membrane lengthening during the extension of the rigid channels in the cytoplasm. This pattern appears to be a compromise between conventional prokaryote and eukaryote chromosome distribution mechanisms in that there is chromosome transport in a passive fashion, but the transport depends on the presence of microtubules. In the dinoflagellates, however, there is no direct connection of chromosomes to the microtubules nor is there formation of microtubules within the nucleus itself. The nuclear membrane remains intact during the division process in dinoflagellates, which is also true for a number of protists and fungi.

Technical difficulties have prevented the detailed analysis of nuclear division events in many of the species with unusual chromosome or mitotic characteristics. If there were more information on a variety of eukaryotic microorganisms we would be better able to determine whether the variations represent sidelines or intermediate stages in the evolutionary sequence of change in chromosome distribution systems. The main point to remember is that an increase in linkage groups becomes possible in eukaryotes because the chromosome transport process functions equally well for a few chromosomes or for a few hundred in any one cell. Lacking such precision, the number of linkage groups must remain low; otherwise genes would be lost and there would be a high mortality rate among progeny cells. The accuracy and reliability of chromosome distribution mechanics provide opportunities for increase in gene numbers, for greatly increased levels of genetic information and potential, and for the complete gene complement to be transmitted to all descendants over countless generations during evolutionary time.

Although both prokaryotes and eukaryotes may generate diversity of genotypes by the processes of mutation and gene recombination, eukaryotes possess an additional means for producing genotype variety in every cell generation. New gene combinations may be produced in eukaryotes through independent assortment of the different chromosomes and of the genes on these chromosomes. Mutation may occur in any cell at any time, involving any of the genes of the complement, and the mutant and wild type alleles of these genes may be transmitted to descendants during asexual or sexual reproductive processes. Recombination of genes, as a function of crossing-over events, also may occur in any cell in which genes from different lineages are brought together, as discussed above. Eukaryotes have a much higher recombination potential than prokaryotes because all the available mechanisms can be used in the generation of genotype diversity. Furthermore, with more than one linkage group it is inevitable that some chromosomes will reassort, leading to recombinants even when no crossing-over has occurred (Fig. 5.5). With recombination by crossing-over between linked genes as an additional parameter, few eukaryotes will fail to produce recombinants. All this is the case for sexually reproducing species far more than for asexual organisms. Since true sexual reproduction does not occur in prokaryotes—that is, there is no fusion of the complete gene sets of two cell lines—these simpler life forms produce fewer recombinants than is theoretically possible for eukaryotes. Relatively few of the highly evolved eukaryotes are asexual, but a considerable number of protists, algae, and fungi have no known sexual stages. Whether these never existed or whether the capacity has been lost in their evolution is uncertain for all these species. But for some it is clear that their ancestors included sexually reproducing types, so the present asexual species must be descendants that have lost the ability to carry on this form of reproduction.

The high premium on diversity in evolution is a function of the display of potentially successful forms in relation to selection pressures. The more diverse the population, the more likely it would be that some one or more of these genotypes would be as successful as, or more successful than, their parent and ancestral predecessors. The limitation on diversity has no long-term evolutionary advantages; therefore mechanisms that ensure diversity would be established rather quickly in populations, once they arose by mutation. One mechanism that ensures greater diversity is sexual reproduction. Since genes from different cell lineages merge and reassort their gene complements prior to transmitting chromosomes to progeny, sexual reproduction provides a regular device for maximizing recombination. Although prokaryotes of some species may introduce some genes from a donor line into a recipient line of cells, only a fraction of the potential recombinants can be produced. In view of the irregularity of prokaryote parasexual mechanisms, an additional decrease in realizing the recombination potential prevails. Of course, asexually reproducing eukaryotes are not much better than prokaryotes at generating recombinants. Moreover, eukaryotes

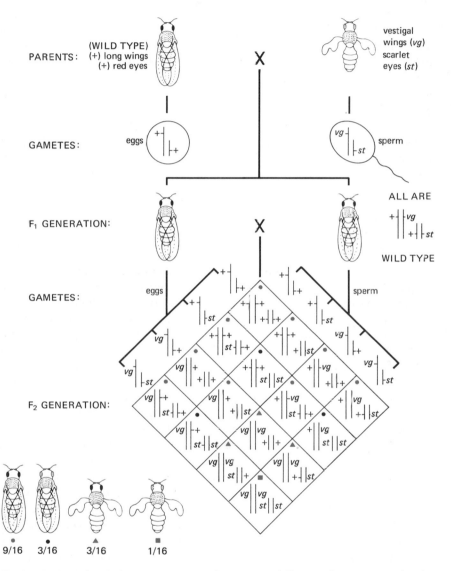

Figure 5.5   Independent assortment of genes on different chromosomes leads to progeny classes that are recombinant as well as others showing the parental phenotypes. The conventional 9 : 3 : 3 : 1 phenotype ratio in the F₂ progeny is indicative of an inheritance pattern in which two different genes on two different chromosomes reassort and the st and vg mutant alleles act as recessives.

which are genetically homozygous are true-breeding whether their reproduction is sexual or asexual. In those true-breeding species there is chromosome assortment and crossing-over within chromosomes, but the products
of these events are genetically identical, resulting in continued homozygosity.

The evolution of sexual mechanisms, to be discussed in greater detail in
another chapter, introduced a considerable selective advantage to eukaryotes. The continued refinement of sexual reproductive characteristics,
such as the separation of the sexes into different individuals, which permitted greater possibilities for different lineages to contribute to the next
generation, is strong evidence in favor of the substantial selection pressures
fixing these traits in evolving species. The series of mechanisms by which
eukaryotes ensure a higher frequency of recombination and therefore
greater genetic diversity explains in large measure their remarkable divergence and success in a relatively brief period of evolutionary time.

Origin of
Eukaryotic Cells

Prokaryotes generally have no persistent internal membranous structures, although a plasmalemma surrounds the entire protoplast. Eukaryotes,
on the other hand, contain a large variety of membranous compartmentations
of the cytoplasmic volume of the cell (Fig. 5.6). They also have a plasmalemma surrounding the protoplast, varying amounts of endoplasmic reticulum, and a membrane surrounding the chromosomes, nucleoli, and other
parts of the nucleus. A number of organelle types are known, but all of
them do not necessarily occur in all eukaryotic cells even of the same individual organism. Single membranes enclose organelles such as lysosomes,
peroxisomes, and various secretion granules. Some subcellular structures
are not bounded by membranes, including ribosomes; the nucleoli within
the nucleus; the chromosomes individually have no surrounding membranes;
and the centrioles or basal bodies are nonmembranous subcellular components. Two kinds of organelles, the mitochondria and the chloroplasts, have
a pair of membranes bounding their contents. Except for simple eukaryotes like the trichomonads, as far as we know all other eukaryotic cells
contain mitochondria. Chloroplasts occur only in the green cells of plants
and in all the algae except for the prokaryotic blue-green species. The
simplicity of subcellular plan in prokaryotes corresponds to their limited
variety of life styles, whereas eukaryotes present enormous diversity at
every level of organization that has been defined, from subcellular to whole
populations.

It is widely accepted that eukaryotes descended from prokaryotic ancestors. Fossil evidence indicates that there were prokaryotic forms more than
3 billion years ago, whereas eukaryotic fossils have been found only in the

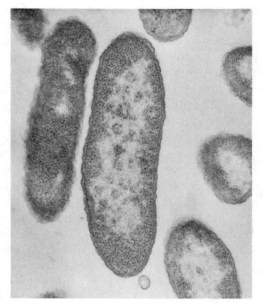

(a)

Figure 5.6 Difference in subcellular complexity is illustrated by comparing the prokaryotic bacterium *Pseudomonas* (a) with the eukaryotic protist *Euglena* (b). In the bacterium the DNA occurs in a less dense central region of the cell, and the only visible membrane is the one that surrounds all the protoplasmic contents; many differentiated structures are present in *Euglena*, including nucleus with chromosomes, chloroplasts, mitochondria, and others. (Electron photomicrographs courtesy of H.-P. Hoffmann)

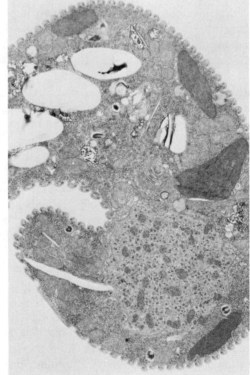

(b)

most recent 1 billion years of deposits. Comparative structure, physiology, biochemistry, and genetics all indicate the derivation of eukaryotes from prokaryotes. Beyond this point there is controversy over the specific phylogenetic lines of descent among the existing species groups. According to the more conventional schemes of classification proposed by some systematists, all present-day groups are derived from prokaryotic photosynthesizing ancestors. Briefly, some authors have proposed that (1) the basic divergence in the present-day world is between plants and animals; (2) photosynthetic eukaryotes (most of the algae and all green plants) descended from a particular ancestral photosynthesizing prokaryote, probably resembling modern blue-green algae; (3) animals and fungi evolved from plants as a function of loss of chloroplasts; and (4) all subcellular structures represent differentiations of the cytoplasm which now typify the eukaryotic cell, such structures including mitochondria, chloroplasts, flagella, the mitotic system, and the endoplasmic reticulum. Variations on this basic proposal have been presented by different authors at different times, but there is little basis in any of these schemes for distinguishing among versions of divergence and pathways leading from prokaryotes to eukaryotic cells and organisms. Intrinsic to all the ideas of this sort that have been presented is that plants and animals represent the significant evolutionary dichotomy. Because of this focus all extant forms have been classified in a manner designed to produce phylogenies consistent with these two important endproducts of evolution, namely, the plant and animal kingdoms. Thus, one is required to include the bacteria in the plant kingdom because they certainly have no animal characteristics; the blue-green algae repose among all the other algal groups even though the former are different in many ways from the eukaryotic algae; fungi are classified as plants although many of their features are more animallike and the group is exceedingly diverse in every respect; and classical arguments rage between botanists and zoologists over whether some unicellular green organisms, such as *Euglena* or *Volvox,* belong in the plant or the animal kingdom.

Another version of phylogenetic divergence, based on the wealth of newer evidence from biochemical, cytological, and genetic studies, seems more acceptable than the other classification schemes, which emphasized the outward appearance or morphology of the species rather than their genetic programs and metabolic capacities. Such a recent version proposed five major groups of organisms: the Monera, which includes all prokaryotic life forms, and four groups of eukaryotes recognized as the Green Plants, Animals, Fungi, and Protists (Fig. 5.7). There are differences of opinion concerning the variety of protists so that one author may place the eukaryotic algae, the simple animals lacking differentiated tissues, and the nonmycelial fungi in the Protista, whereas another author will group these three types among the Plants, Animals, and Fungi, respectively. These two authors would agree that euglenoids, protozoans, and a host of simple eukaryotic life forms lacking complex tissues belong in the Protista. They also would agree on the species to be classified as Monera, Plants, Animals, and Fungi, except for the disputed protist or protistlike forms. The fundamental agreements are significant, and the essential dichotomy is that of

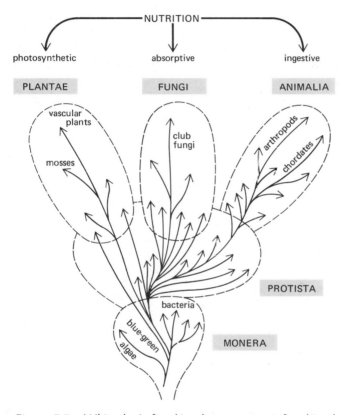

Figure 5.7   Whittaker's five-kingdom system. A five-kingdom system of classification based on levels of organization: the prokaryotic Monera; eukaryotic Protista, which lack tissue-forming capacity; and the eukaryotic Plantae, Fungi, and Animalia, which are highly diverse in their multinucleate, multicellular patterns of development. The nutritional modes are quite varied in the protists, whereas most plants are photosynthetic, fungi are absorptive, and animals principally ingest foods. Most of the Monera are absorptive, a few are photosynthetic, and none are ingestive. (Adapted from "New Concepts of Kingdoms of Organisms" by R. H. Whittaker, Science, Vol. 163, pp. 150–160, Fig. 3, 10 January 1969. Copyright © 1969 by the American Association for the Advancement of Science. Reprinted by permission of R. H. Whittaker and the American Association for the Advancement of Science.)

prokaryotes and eukaryotes, not plants and animals, as the primary divergence.

Lynn Margulis recently has revived a theory of eukaryotic cell origin, which has been accepted by an increasing number of biologists in the few years since 1967 when she first presented newer lines of evidence and argument in support of the older proposition. In essence she has revived the proposal that eukaryotes evolved from prokaryotes by a sequence of specific endosymbioses (Fig. 5.8). The particular components of the eukaryotic cell assumed to be modern remnants of ancient symbiotic prokaryotes include the chloroplast, the mitochondrion, and the eukaryotic flagellum. The suggested sequence of events as proposed in the Endosymbiosis Theory is as follows:

1. Ancestral prokaryotes diversified into the numerous kinds of Monera, which included heterotrophic and photoautotrophic forms.

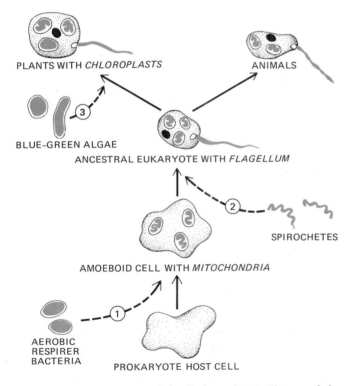

Figure 5.8 A summary of the Endosymbiosis Theory of the origin of eukaryotic cells, as proposed by L. Margulis. (Adapted from *Origin of Eukaryotic Cells* by L. Margulis, Yale University Press, p. 58, 1970. Copyright © 1970 by Yale University. Reprinted by permission of Yale University Press.)

2. An amoeboid prokaryote engulfed a symbiotic aerobic respiring pro-karyotic organism, and these respiring symbionts persist in modified form as the mitochondria of present-day eukaryotic cells.

3. Such an amoeboid ancestor established a symbiotic relationship with spirochete types of prokaryotes, thus acquiring a motility mechanism, which ultimately evolved into the present-day mitotic mechanism and the flagellar apparatus, both of which have microtubule components.

4. Incorporation of photosynthetic prokaryotes, such as blue-green algae or photosynthetic bacteria, occurred in some of the newly motile, newly respiring protoeukaryotes, thus giving rise to the ancestors of the eukaryotic algae and the complex green plants.

5. Differentiation of a nuclear membrane, a mitotic mechanism for accurate distribution of the genetic material, and multiple linkage groups (becoming chromosomes) occurred.

All these events plus others are proposed as steps leading eventually to eukaryotic cells capable of motility and aerobic respiration. Only those protoeukaryotes that happened to incorporate photosynthetic symbionts became photosynthesizing, themselves; all other protoeukaryotes evolved into nongreen heterotrophs such as the protozoans. Animals and fungi are proposed to be descendants of heterotrophic eukaryote ancestors, rather than forms derived from a photosynthetic ancestor by virtue of plastid loss.

This classification scheme is satisfying in many ways, especially in pro-viding a reasonable set of evolutionary lines of descent based on the pro-karyote–eukaryote dichotomy. Furthermore, the suggestions permit a phy-logenetic distinction among plants, protists, fungi, and bacteria according to biochemical, cytological, and genetic lines of evidence rather than the more superficial and limited criteria of external appearance and simple cell plan. On the older classification plans many protists and all the bacteria and fungi usually were inserted into the Plant Kingdom, particularly because of the rigid cell wall and wall chemistry.

While there are numerous satisfying aspects to schemes of classification based on the prokaryote–eukaryote dichotomy, there are some portions of the theory that may be open to alternative explanation. In particular, origin of eukaryotic cells from prokaryotes via symbiont incorporations may apply to one of the pertinent organelles and not to others. The weakest lines of evidence are those concerning the evolution of the microtubular components of the cell and, especially, the evolution of the flagellum from a prokaryotic spirochete. Much of the evidence presented in support of this notion is vague and scattered, and dependent in large part on comparison with some peculiar modern protozoa that do utilize spirochetes in their motility mechanism. Such spirochete-motility examples are quite infre-quent, and it seems that the comparison has been stretched beyond capacity to make fit with a hypothetical evolutionary event. The ubiquity of microtubules in eukaryotic cells and their participation in a variety of subcellular movement phenomena, including cytoplasmic streaming among others, argues for a more fundamental set of evolutionary modifications than the incorporation and subsequent change in a symbiotic spirochete population. It is important to realize that the basic floor plan of nine microtubule doublets encircling a pair of single microtubules (the "9 + 2" arrangement) in cilia and flagella of all eukaryotic cells implies an ancient ancestry and a common set of evolutionary selection pressures. Furthermore, not all basal bodies (centrioles, when present in the mitotic apparatus), which subtend the flagella and cilia, appear to contain DNA. The presence of this presumptive genetic material contributes substantially to the argument Margulis presents for endosymbiosis leading to flagella from spirochete precursors.

A compelling basis to the recent acceptance of the Endosymbiosis Theory is the occurrence of a functioning genetic appartus in mitochondria and chloroplasts. There is no evidence at all to indicate that centrioles, or basal bodies, also contain a genetic mechanism. For this reason, too, we ought to divorce the flagellum from our consideration of the ancestry of mitochondria and chloroplasts, the other two candidates for having a symbiotic past. In mitochondria and chloroplasts there is DNA that resembles prokaryotic DNA in being duplex naked molecules unassociated with proteins (Fig. 5.9); there is a protein-synthesizing apparatus which includes ribosomes and the required functional RNAs such as ribosomal, transfer,

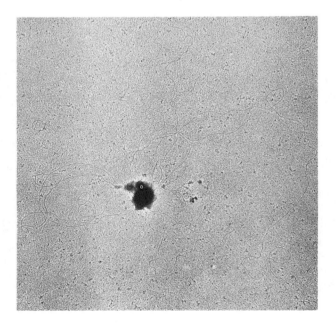

Figure 5.9 Released duplex DNA from a mitochondrion of a yeast cell after exposure to osmotic shock conditions. Compare with Fig. 5.1. (Photograph by the author.)

and probably messenger species; and specific drugs that affect macro-molecular syntheses in prokaryotes induce similar responses in the organelle systems while having no effect on the macromolecular syntheses in the nucleocytoplasmic systems of the very same eukaryotic cells. There is an increasing body of evidence showing that organelle DNA is informational, although a considerable number of nuclear genes code for organelle-specific molecules and functions in addition to any genes that occur in organelle DNA itself. Some biologists reject the possibility of the development of a separate genetic apparatus in the organelles alongside the major nuclear genetic system, and they find the Endosymbiosis Theory a satisfying ex-planation for organelle genetic systems. Quite simply, these are assumed to be remnants of the endosymbiont genetic appartus which have been re-tained because of selective advantages. The nature of such selective ad-vantage remains unspecified on either the ancient symbiont theory or alter-native explanations involving differentiation of an indigenous organelle genetic system.

As more information becomes available to unravel the nature and variety of organelle genetic systems it becomes more evident that there are traits that have been ignored or deemphasized in the discussions. It seems very clear that metazoan mitochondria contain very similar genetic components, especially having circular duplex DNA molecules measuring about 4.6 to 5.9 micrometers (μm) in contour length; small ribosomes having sedimen-tation values of 55 to 60 S; and information that is transcribed primarily as ribosomal RNA. Protozoans, the presumptive ancestors of the metazoans, have been studied very little for these organelle traits but the available data show clearly that the duplex DNA molecules may exist either in circular or linear conformations and may be as small as 0.5 μm or as long as 17.6 μm; and that the mitochondrial ribosomes of at least one ciliated protozoan are about the same size (80 S) as the cytoplasmic particles in the cell. There have been reports for very few species of fungi, but all appear to contain ribosomes that are larger than those in metazoan mitochondria. Also, their DNA molecules generally are longer and may even exist in circles as small as 2 μm as well as in a heterogeneous display of contour lengths up to about 25 or 26 μm. The actual lengths and conformation of DNA from mitochondria of green protists like *Euglena* and flowering plants such as spinach and corn and tobacco also are controversial. While some laboratories have reported the occurrence of circular DNA 30 μm long, others have found only longer and shorter linear molecules or very small circles.

Although earlier studies were inconclusive concerning the conformation of chloroplast DNA molecules, recent reports have provided good evidence for the presence of circular DNA measuring 40 to 45 μm in contour length. This indicates a substantial difference between the potential information content of mitochondrial and chloroplast DNA even in the same species, as well as striking differences between different species in their organelle genetic materials. The significance of these differences remains to be established.

There is little basis for considering all mitochondrial genetic systems to be essentially alike or to be so different from the genetic appartus of the remainder of the cell that we ought to invoke the Endosymbiosis Theory and claim that the problem of organelle origin is solved. We should ask how the mitochondrial genetic system originated and how it evolved if not by the proposed mechanism of endosymbiosis. There is no satisfactory explanation and no particular evidence to support any explanation we might suggest. Perhaps the mitochondrion did differentiate from some portion of the ancestral prokaryotic cell, as has been suggested for the nucleus and as we know to be the case for such modern membranous organelles as lysosomes and peroxisomes. The crucial requirement for efficient systems of aerobic respiration, and the additional requirement for a greater volume devoted to respiratory activities than could be achieved with the respiratory chain localized in the plasmalemma, would have conferred great selective advantages favoring any mutants that could improve in this way. There may be particular advantages that would explain the segregation of some genes into the mitochondrion while others were retained in the nucleus, but since we know so little about mitochondrial gene complements and the unique features of other parts of the organelle genetic system, it may be fruitless to speculate.

One persistent and nagging question concerns the striking similarities between prokaryote DNA and mitochondrial DNA, and the similarity of response to a spectrum of drugs that have been tested. Common ancestry and relationships by descent have been proposed by way of explanation according to the Endosymbiosis Theory. But some recent evidence leads one to question the reality of these similarities in response to drugs. There are mutants whose mitochondria are resistant to drugs such as chloramphenicol and erythromycin, whereas the wild type mitochondria are sensitive to these antibiotics. The effect of these chemicals is expressed in the mitochondrial protein-synthesizing system, since all known protein synthesis on cytoplasmic ribosomes appears to be resistant to chloramphenicol and erythromycin. The discovery of strains of species that have become genetically resistant in their mitochondrial compartment raises the possibility that we have not yet found mutant strains in which the cytoplasmic system has become sensitive to both drugs as a consequence of nuclear gene mutations.

What we have assumed to be a sharp division between mitochondrial and nucleocytoplasmic system responses to various antibiotics, and have used as an evolutionary criterion, may be only a reflection of varying allele frequencies in the two genetic compartments of the cell. Theoretically each gene may exist in a series of allelic forms, that is, as two or more alleles per gene. Populations differ in frequencies of allelic forms of a common set of genes, and it is not unusual to find one population with a preponderance of one allele while another population of the species contains individuals with the alternative allele(s) in high frequency. We would not conclude that different populations belonged to different species or phyla merely because of varied frequencies for alleles of a few genes. But we seem to do just this in emphasizing the different responses of mitochondrial and nucleo-cytoplasmic systems to chloramphenicol, erythromycin, cycloheximide, and a relatively limited group of chemicals.

Whether or not allele frequencies provide an acceptable alternative explanation for the observed dichotomy of response to chemical agents, we still are faced with the question of similarities between organelle and prokaryote DNA molecular packaging. One possible explanation could be that the very same sorts of selection pressures that contributed to prokaryote DNA genophores also led to the striking resemblance to organelle genophores. There would be similar problems of accurate and reliable distribution of a complete set of genes to new mitochondria because there is no available mitotic mechanism. There would be similar advantages to a single linkage group, which would provide greater chance for distribution of a complete gene set to the new mitochondria, as occurs in prokaryotes. Of course, there might be more than one linkage group per mitochondrion or more than one genetic species of mitochondrion in a cell population. It is generally agreed that multiple copies of DNA exist in mitochondria of many species; therefore some loss could be tolerated in new mitochondria because the information would be present in those DNA strands that did find their way into newly formed structures.

The mitochondrion is not dependent entirely on its own DNA complement: in yeast there is a class of mutants with no mitochondrial DNA at all that consistently produce new mitochondria in each generation. These DNA-less mitochondria, however, are functionally defective and morphologically different from wild type organelles in that the inner membrane produces no infolded cristae and aerobic respiratory activities are absent.

These suggestions require substantiation by experimental methods or by analysis of mitochondria from many species of varied evolutionary rankings. In view of there being no direct evidence in support of endosymbioses either, we would be wise to keep our options open and not foreclose on the interesting possibility that eukaryotic cells contain at least two genetic systems, which have evolved jointly. If each system provides some unique contribution to the total cellular economy, then selection pressures would operate to retain both systems in all eukaryotic cells.

For chloroplasts the case may be quite different from that for mitochondria; that is, the evidence for chloroplasts as ancient symbionts is somewhat more compelling than it is for mitochondria. There is more DNA in a chloroplast; there is an extremely active protein-synthesizing system with abundant ribosomes present (Fig. 5.10); and chloroplasts have been incorporated into animal cells where they have survived intact and been recovered undamaged after a period of residence. It is unlikely that animal cells harbor any nuclear genes specific for chloroplast development and function, so this organelle seems to be more autonomous than the mitochondrion. There also have been reports of chloroplast reproduction in vitro (outside the cell in an artificial environment), which implies the capability for some level of independent existence. Many green cells certainly can survive loss of their chloroplasts because there is a heterotrophic metabolism that can provide the necessary energy and carbon for life. No cell can survive the total loss of its mitochondrial complement, however; the remaining anaerobic metabolic equipment is insufficient to maintain the vast majority of eukaryotic cells for all their energy and life requirements. Because all eukaryotic cells contain mitochondria it has not yet been possible to incorporate mitochondria from one species into the cells of another by usual methods. But using the technique of hybridizing somatic cells in

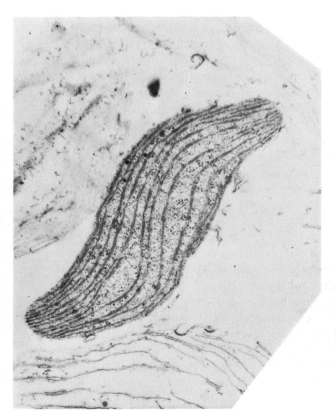

Figure 5.10 Thin section through a chloroplast of *Euglena gracilis*, photographed using the electron microscope, showing numerous ribosome particles in the matrix region of the organelle. (Photograph courtesy of H.-P. Hoffmann.)

culture it has been shown that the mitochondrial population that persists is the one associated with the cell type whose chromosomes are retained in the hybrid cells. This new approach may make it possible to pursue questions about mitochondrial control over its own characteristics versus control exerted by nuclear systems. Since yeast mutants lacking mitochondrial DNA still can produce mitochondrial structures with two surrounding membranes, it is obvious that nuclear genes exert a considerable level of control over the formation of the organelle.

Finally, we really do not know how new mitochondria form. A few experimental studies have reported data interpreted to mean that new mitochondria form from preexisting mitochondria. This mechanism of mitochondrial biogenesis would fit neatly into the Endosymbiosis Theory since it invokes a sort of binary fission process such as occurs in prokaryotes. It could fit equally well into alternative ideas, especially the one that the similarities are fortuitous and merely reflect the action of similar selection pressures on different genetic lineages. But before such explanations or suggestions are made we must obtain unequivocal evidence concerning the mode of mitochondrial biogenesis, and such evidence still is lacking.

The basic difference between the evolutionary pathways stipulated by the Endosymbiosis Theory and the alternatives that have been proposed here is the distinction between parallel and convergent evolution, or homology and analogy. In examples of parallel evolution we find that divergent lineages experience similar mutations and continue to evolve along separate but parallel pathways, thus producing descendant lines that resemble each other. This is the evolutionary phenomenon proposed by the Endosymbiosis Theory—that is, that the ancient prokaryotic symbiont and the modern mitochondrion and chloroplast have retained their fundamental similarities, by virtue of parallel evolutionary modifications, for at least the past billion years or more. There are no known case histories showing such levels of conservatism and such durations of parallel evolution, but that does not mean it is impossible. The alternative suggestion, that the organelles may have evolved as differentiations of the cytoplasm and that the apparent similarities reflect similar selection pressures acting on different gene complements in organelles and prokaryotes, is based on the proposition that convergent evolution has occurred. Convergence leads to apparent similarities in traits that have arisen from genetic changes in different information sets; that is, the resemblances are superficial and have been produced as a consequence of mutations in completely different genes.

An example of convergent evolution would be the eye of vertebrates as compared with the eye of an invertebrate such as the octopus or squid. There are many apparent similarities, but close study of the development of the organs reveals that quite different origins, tissues, and processes led to the final appearance of genetically different structures that are very similar functionally. In another case of convergent evolution, the wings of in-

sects and birds are similar in function and have some common features, but their developmental origins are exceedingly different, and, again, different sets of genetic information led to superficially similar structures. While it is not altogether certain in every case, many of the marsupial mammals of Australia and the placental mammals elsewhere in the world evolved along very similar lines. Such convergent evolutionary pathways produced marsupial equivalents of the placental moles, squirrels, various predators, and other kinds of superficially similar species.

Considering all that we still have to learn about organelles, it is worthwhile to view the problems of organelle evolution as still open to further investigation. It may be easier to discard mitochondria as ancient symbionts, but we may be discarding an interesting system of eukaryote genetic evolution. If there is some possibility that mitochondria evolved as subcellular differentiations in eukaryotes and not as alien forms that took up residence in a compliant host cell, then it would be eminently worthwhile to keep the controversy alive in the hope of finding the answers in the near future to the questions we pose today.

Some Suggested Readings

Avers, C. J., F. E. Billheimer, H.-P. Hoffmann, and R. M. Pauli, "Circularity of yeast mitochondrial DNA." *Proceedings of the National Academy of Sciences* 61 (1968), 90.

Barghoorn, E. S., "The oldest fossils." *Scientific American* 224 (May, 1971), 30.

Margulis, L., "Symbiosis and evolution." *Scientific American* 225 (August, 1971), 48.

Pichichero, M. E., and C. J. Avers, "The evolution of cellular movement in eukaryotes: The role of microfilaments and microtubules." *Sub-Cellular Biochemistry* 2 (1973), 97.

Raff, R. A., and H. R. Mahler, "The nonsymbiotic origin of mitochondria." *Science* 177 (1972), 575.

Stanier, R. Y., "Some aspects of the biology of cells and their possible evolutionary significance," in *Organization and Control in Prokaryotic and Eukaryotic Cells* (ed., H. P. Charles and B. C. J. G. Knight). London, Cambridge University Press, 1970, pp. 1–38.

Whittaker, R. H., "New concepts of kingdoms of organisms." *Science* 163 (1969), 150.

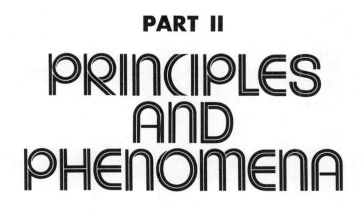

PART II

PRINCIPLES AND PHENOMENA

CHAPTER 6
MECHANISMS THAT PRODUCE DIVERSITY

A basic observation of life on the Earth is that it is exceedingly diverse. Whether one observes life from vast areas of the planet or even the organisms contained in a pinch of garden soil, a fantastic variety of living types can be found. Different ideas have been presented throughout history to account for this living variety, including special creation and immigration from elsewhere in the universe. These two particular concepts actually avoid the basic issue on the one hand and do not lend themselves to experimental verification on the other. Furthermore, there is very little evidence to support either idea, although one does not require evidence to accept an article of religious faith.

The evidence used to support the notion of life's arriving from other parts of the universe exists principally in the contents of some kinds of meteorites. The presence of hydrocarbons and other organic chemicals may be indications of synthesis by life forms elsewhere than on Earth, although many such molecules have been produced under abiogenic conditions in laboratories on our planet. Some polypeptide molecules have been analyzed from meteorite samples and those containing amino acids unlike those produced by our own kinds of life may have been produced either by extraterrestrial life or abiogenically. Various kinds of plant spores have been isolated from meteorites, but in most cases these probably were picked up by the rocks during their travels through our atmosphere rather than having been carried as baggage from the beginning of the trip to its termination somewhere on the Earth's surface. Certainly, ragweed pollen could exist elsewhere in the solar system, but it seems an unlikely explanation for its presence in an alien rock sample. Since we consider it very likely that life originated on the Earth under suitable conditions, there may be life elsewhere in the universe as well as on the Earth. Perhaps some evidence of extraterrestrial life has been discovered in some of these rocks, but this evidence produces a rather weak line of argument in support of all life having come from non-Earth regions in the beginning and then having changed with time to the spectrum of species seen in the fossil record and in present-day habitats.

The first historically documented theory proposed to explain the origin of diversity was the theory of Jean Lamarck, published in 1809. The theory is known by various descriptive epithets, including the Theory of Use and Disuse and the Theory of Inheritance of Acquired Characteristics. Although we have rejected the specific premises stated by Lamarck and others who expanded on his ideas, the presentation of the theory was historically important as the first attempt to support the suggestions based on a thorough and systematic collection of observations and of interpretations based upon these observations. Lamarck attempted to develop some common basis by which all observed diversity could be explained, and his ideas have such compellingly optimistic overtones that they crop up repeatedly even in modern writings. In essence his theory stated that the environment, or certain environmental conditions, caused the specific changes in the organism; that is, the environment provided the stimuli and changes in the organism were direct responses to these stimuli. After a change had been initiated, Lamarck proposed that the altered trait continued to be inherited even if the stimulus was removed at some later time. All changes induced in response to stimuli were considered to be adaptive so that change always was in the direction of increased benefit to the organism. The ultimate species was considered to be our own, toward which all other species were supposedly evolving.

One often-quoted example from Lamarck's writings concerns the change in neck length of the giraffe. On the basis of Lamarck's theory the ancestral short-necked giraffe browsed on low vegetation. As this food source became depleted, the giraffe stretched toward the higher leaves of trees, and with continual stretching the neck became longer until it reached its present length. Even though browse food again became available at lower as well as higher reaches, the induced change became an inherited trait and was transmitted to descendant generations. The change was beneficial because the giraffe could secure its food despite environmental hostilities of past times. While this may seem a little silly to us now, there are many examples of explanations of modifications in life forms in Lamarckian terms. For instance, housefly populations have become increasingly resistant to DDT chemical sprays. If one explains this change as a necessary event for the houseflies to survive the insecticide exposure, then the explanation is Lamarckian in the sense that it implies that the change to resistance was an adaptive response to the new environmental conditions. In the same way, we know that more and more harmful bacteria have become resistant to the effects of antibiotics. Why? Lamarck would have said that the change was induced by the needs of the bacteria to survive in an increasingly hostile environment; that is, the characteristic of antibiotic resistance was acquired by bacteria exposed to the drugs, and this change subsequently was inherited by descendants who continued to be resistant even when drugs no longer were present. These particular phrasings are used by many people to explain various examples of inherited diversity, but each explanation is contrary to everything we know about genetic material today.

In effect, Lamarck's theory proposed the phenomenon of directed mutation. There is absolutely no evidence from the past 75 years of genetic studies to indicate that any gene will mutate in response to the need for survival of any living species. If we consider Lamarckian doctrine as a proposal of specific directed mutation, then the entire theory must be rejected for lack of evidence. Of course, many independent lines of evidence refute Lamarckian explanations, but we often are seduced into using such inaccurate phraseologies. Sometimes it is an innocent misstatement; in some cases the entire basis for the explanation is false. It apparently is very difficult to reject entirely a theory based on the optimistic premise that all change predictably will be beneficial to the species. For adherents of Lamarck the future is bright with promise because only adaptive mutations will occur. Unfortunately there is no support for such a future prospect if its basis is directed mutation.

Heredity was a mysterious and little-understood branch of knowledge in Lamarck's time, and his theory was acceptable as long as there was no particular basis for considering it to be wrong. Darwin incorporated some of Lamarck's theory to explain the origin of diversity, but he was more inclined to believe that other mechanisms might be operative even though nothing was known of such mechanisms when *The Origin of Species* was published in 1859. Darwin actually was less interested in the origin of diversity and more concerned with patterns of change in species and with explanations for preservation, extinction, and improvement in species groups. One problem that plagued Darwin was the idea current at that time that diversity would be decreased by interbreeding between inherently different forms. Many examples of apparent blending inheritance patterns were known in the mid-nineteenth century, such as that crosses between different parents usually resulted in progeny that were intermediate in appearance relative to the two parents as well as being fairly similar to one another. Inheritance patterns of this kind were explained easily after 1900 when Mendel's genetic studies were rediscovered but were not understood during Darwin's time. Despite the problems in understanding heredity, Darwin's Theory of Natural Selection was a very sound treatise and has easily accommodated newly discovered genetic and evolutionary principles, to the present day.

Mutation

The first modern proposal to explain the origin of variation was the Mutation Theory of Hugo De Vries, a Dutch botanist who was one of the three rediscoverers of Mendel's 1865 genetics studies. De Vries's book was published in 1901 and presented a substantial body of evidence in support of the concept that inherited variation arose suddenly and randomly. These sudden heritable changes, which he termed mutations, sometimes produced minor changes in an organism and sometimes led to substantial modification in the mutant. The term and concept have remained in use to the present day, but, curiously, many of the mutants studied by De Vries were individuals that had undergone drastic modifications of their chromosome complement rather than small changes in one gene. De Vries worked with the flowering plants of the genus *Oenothera,* the evening primrose, and most *Oenothera* species have very unusual chromosome arrangements. The peculiar chromosome complements of *Oenothera* led to many confusing genetic results, but such unusual plants provided the basis for experimental conclusions that we hold valid in general principle to the present day.

Mutation is the only known mechanism by which *new* genetic information arises in all life forms. Most mutations involve relatively small changes in the informational content of the gene, whether among the codons of DNA in most organisms or in RNA codons of viruses lacking a DNA-based genetic system. There are excellent studies showing clearly that nucleic acid modifications may involve as simple an event as the substitution of a single nucleotide in one codon of a gene. Such a substitution may lead to a different amino acid in a specific site in the polypeptide chain of a hundred or more of these residues and be enough to alter the appearance (phenotype) of the mutant cell or individual. Addition or deletion of a nucleotide in a codon results in an altered reading frame, leading to a modified protein, which may be completely nonfunctional. Such frameshift mutations have been demonstrated in a variety of species, as have substitutions and other specific changes in the genetic information encoded in the DNA molecules.

One of the earlier lines of evidence in support of the idea that an altered amino acid reflected a mutational change in the gene that specified the protein was derived from studies of the beta chain of human hemoglobin. People with sickle-cell anemia have an aberrant hemoglobin, and pedigree analysis reveals that this protein difference is due to a pair of alleles of one gene. In the earlier studies the protein was digested and the fragments of the digest were "fingerprinted" by the method of paper chromatography. The 26 peptide fragments from normal and aberrant hemoglobin could be compared by examining the positions occupied on the paper by each of these parts of the digested molecules. One of the 26 spots was different; when this peptide fragment was analyzed for its kinds and sequence of amino acids it was discovered that the amino acid valine occurred in aberrant hemoglobin S in the place occupied by glutamic acid in normal hemoglobin A. Another mutant allele of this gene is responsible for the hemoglobin C disease, and analysis of this protein variant showed that an amino acid substitution existed in exactly the same site in the protein molecule, but lysine was the replacement at this location in hemoglobin C. Since these three proteins are products of action of three different allelic forms of the same gene, the most probable explanation was that the same codon was altered in both mutants but in different ways. Each altered codon led to a different amino acid substitution as a consequence of the mutational change in the gene (Fig. 6.1).

With specific information from later studies of the genetic code it was possible to deduce the probable mutational change that led to each hemoglobin variant. The codons that specify glutamic acid in the normal hemoglobin A (Hb A) individuals are GAA and GAG. For valine to be substituted, as in hemoglobin S (Hb S), the mutation may cause a change from GAA to GUA or from GAG to GUG. In either case the middle nucleotide in the triplet codon is modified. Hemoglobin C (Hb C) variants contain a lysine residue instead of glutamic acid at position number 6 of the beta chain of the hemoglobin protein, and this substitution may result from a change of GAA to AAA or from GAG to AAG. In this case the substitution has occurred in the first nucleotide in the codon. Since these codons

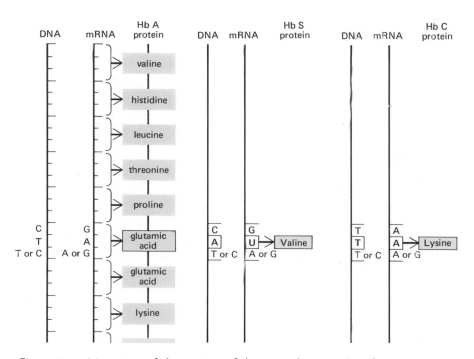

Figure 6.1   Mutations of the portion of the gene that specifies the amino acid composition of the beta-chain of human hemoglobin. A substitution of the middle base (T → A) leads to the altered Hb S protein, whereas a substitution of cytosine by thymine in the first base of the triplet at position number 6 in the gene yields the mutant Hb C protein. The remaining 103 amino acids are identical in all three hemoglobin types.

are transcripts of DNA information and are complementary to the nucleotides in this region of the gene, the actual mutations in the DNA molecule involved the substitution of adenine for thymine in Hb S and of thymine for cytosine in Hb C variants (see Fig. 6.1). Complementarity underlies the interactions between nucleic acid molecules in the genetic system, so we can make these assumptions without resorting to direct analysis of the separate elements. This simple change in one nucleotide of one codon in the 104 contained in the gene for the beta chain of the hemoglobin molecule has severe consequences for those humans who have received a mutant allele from each parent and are, hence, homozygous for the trait. Sickle-cell anemia leads to various physiological and clinical damage, and often causes premature death of the individual.

In similar studies, proteins that have been analyzed for kinds and sequence of amino acids, either in the entire protein or only in the modified portion, also have been shown to contain altered regions coincident with gene mutations. The general principle that emerges is that the gene and the protein are co-linear. A change at any one location in the gene leads to a specific change in the protein in precisely the location specified by the codon that is translated into the particular amino acid for which it codes.

Direct evidence for co-linearity of gene and protein has come from the elegant studies by Charles Yanofsky of the tryptophan synthetase gene and protein system in *Escherichia coli* (Fig. 6.2). The normal or wild type gene in this bacterial species specifies the enzyme protein tryptophan synthetase containing the amino acid glycine in position number 210 of the sequence. The codon that is translated as glycine is GGA, and mutant strain A23 contains the RNA codon *A*GA (specifying arginine) while mutant strain A46 contains the RNA codon G*A*A, which translates as glutamic acid. If mutants A23 and A46 are crossed, then one of the recombinant products should be the wild type with codon GGA and with glycine in position number 210 in the enzyme protein molecule. The experimental evidence shows that a mutant may arise from as simple a change as one nucleotide in one codon, that different mutants may occur as a consequence of a substitution of any one of the nucleotides of a triplet codon, and that the information in the nucleic acid of the gene is translated into the amino acids of the protein at sites that are co-linear. In addition to this particular set of experiments, which demonstrated co-linearity of the gene and the protein product of gene action, other experimental analyses have provided supporting data to substantiate this general genetic principle.

Mutations are random events. Any gene may change at any time in any cell of a population; thus the occurrence of inherited variation is a function of chance. But while the individual mutational event is unpredictable as to time and place of its appearance, all genes are susceptible to mutation because of the common features of nucleic acid chemistry. It follows from

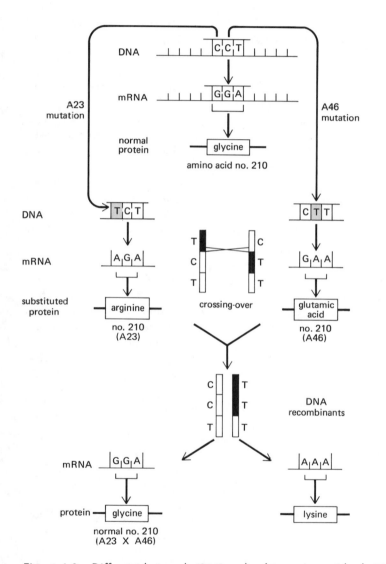

Figure 6.2   Different base substitutions lead to amino acid substitutions in the tryptophan synthetase protein in *Escherichia coli*. Crosses between the A23 mutant and A46 mutant produce recombinant progeny, some of which have the wild type protein. Recombination verified the co-linearity of the gene and protein molecules.

this that, under standard conditions, each gene has a particular probability for change, which is termed its *mutation rate*. Because there is a constant probability for every gene to mutate, it is possible to make mathematical predictions for any population that a constant proportion will be mutant for a particular gene. Such predictions relate only to the event and have no bearing whatsoever on the relative benefit or harm to the population as a consequence of mutational changes. The direction or benefit of a gene modification is entirely unpredictable. Statistically, however, it is more likely that a mutation will be detrimental to some degree than that the alteration of the gene will lead to some improvement for the population. Any random change in an efficient system that has undergone a long history of stepwise improvements is more likely to disrupt the activity of a complex mechanism in that system. To use an analogy, if we take a wrist-watch and tap it on a hard surface such as a table or a desk, it is more probable that we will damage the mechanism than that we will improve it, because any one of a number of changes may lead to disruption of its intricately coordinated operations. On the other hand, because the change will be random, there is a low probability that an improvement might occur. In the example of the watch, we may luckily dislodge some dust particle that hampered the mechanism and thus recover a more accurate timepiece than we had before the tapping began.

A large body of experimental evidence indicates that mutation rates vary for different genes, and even for different sites within the same gene. This is to be expected; genes vary in nucleotide composition and sequence, and some mutational events are more likely to occur in one part of a sequence than in another because of chemical differences intrinsic to particular nucleotide combinations. Seymour Benzer's analysis of the *r*II gene (cistron) in the T4 bacteriophage demonstrated the occurrence of "hot spots," that is, of particular mutational sites that recurred frequently in experiments. Other locations in this cistron mutated less often or not at all during the course of this extensive investigation. Studies such as these also indicate that one gene may exist in a large number of allelic forms, since each unique modification results in a uniquely new nucleotide sequence for the same gene. Many allelic variations remain undetected in populations because not every mutation leads to a recognizable phenotypic change in the organism.

That most mutations are harmful to some degree was shown clearly by H. J. Muller in his studies of the common fruit fly *Drosophila melanogaster*. The experimental design he devised has been modified somewhat in recent years, but the underlying principle is valid and useful in many genetic analyses. Muller used special stocks of *Drosophila* with marker loci on the X chromosome; *C* corresponded to a region that reduced or prevented crossing-over, so that the chromosome was transmitted intact and could be followed as such for a number of generations; *l* was a locus that was lethal to the organism if present in double dose but masked in the presence of the wild type allele L; and B referred to the bar-eye mutant allele which acted as a dominant in the presence of the wild type allele b for normal numbers of eye facets. These ClB stocks furnished heterozygous females for the experimental crosses with wild type males, whose genotype could be described as cLB, or by the more conventional wild type symbols $+++$. All these genes are located on the X chromosome, whether of males that had been irradiated or of untreated members of laboratory or natural populations.

In experiments designed to detect sex-linked mutations (Fig. 6.3), the initial crosses yielded three surviving genotypes in the F_1 progeny. F_1 females that were visibly bar-eyed, and therefore carried the marked ClB chromosome, were then mated with F_1 wild type males. Examination of the F_2 progeny classes yielded the desired information. Among these F_2 there were expected to be four genotypes, each phenotypically recognizable: (1) ClB females, (2) wild type females, (3) wild type males, and (4) ClB males, which would not be recovered since each possessed the lethal allele on the X chromosome and no masking allele occurs on the Y chromosome. Surviving progeny therefore were expected to consist of twice as many females as males because half the males die in the F_2 experimental generation. From counts of males and females in progenies of separate crosses, and from examination of individual males, it was possible to discern what kinds of sex-linked mutations had been present in the X chromosome of individual males used as parent in each original cross. Any visible mutations, such as modifications of wings, eyes, or bristles, would be obvious on simple inspection of the F_2 males from every set of crosses. If a lethal mutation had been present or had been induced by x-ray exposure in some experiments, then no males would be present among the F_2 progeny of a cross that had involved that particular male parent individual. One very important feature of this experimental design is that detrimental mutations also could be detected on the basis of the proportion of surviving males in the F_2 population, even if there were no visible signs of mutation. Thus mutations conferring some degree of harm to the population would be evidenced by a decrease in numbers of males expected, although not to the zero level of the lethal mutant type. If there were twice as many females as males, then no detrimental mutation had occurred on the original male parent's X chromosome. Any decrease in numbers of males from the one-third expected in the population, but not to zero, indicated the existence of a detrimental gene mutation.

The most frequent mutations found in these experiments were detrimental, and the least frequent types produced visible morphological changes in the mutants. Lethals occurred in an intermediate frequency relative to the other two classes of mutation. Detrimental mutations therefore are the commonest type and may affect populations in any number of subtle ways.

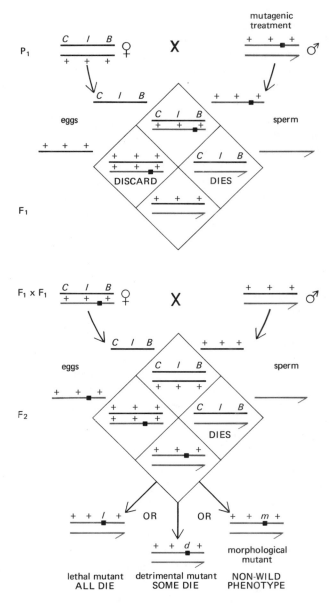

Figure 6.3   A diagram showing the classical *CIB* technique first used by
H. J. Muller to demonstrate the occurrence of sex-linked, recessive mutations
in *Drosophila*. The proportions of males in the F₂ progeny (from zero to
one-third) and their appearance served as diagnostic clues to the nature of the
mutational consequence in the population.

There may be a shorter life span, greater susceptibility to infection, less than optimum adult size, or any sort of manifestation of lowered fitness of the population. Such hidden mutant alleles constitute the "genetic load" of a species, and every outbreeding species carries large numbers of recessive mutant alleles that are potentially harmful. In a later chapter we will consider some aspects of the gene pool of populations and means of estimating the genetic load and its significance in evolution.

There is of course ample experimental evidence of beneficial mutational changes as well as of detrimental effects of gene modifications. Such evidence shows clearly that beneficial mutations are very low in frequency as compared with harmful modifications. From earlier studies of barley plants we have accepted a general frequency of beneficial mutations to be about 1 per 1000. Those experiments showed that mutants induced by x-irradiation (x-ray–induced and spontaneous mutation are entirely comparable except for the increased rates of induced mutation proportionately for all the genes of the complement) were almost always less successful than wild type individuals. But, on the average, one in every thousand mutants showed a noticeable improvement: increased tolerance to drought, higher yield of seeds and fruits, resistance to infections, and various other traits that could be documented as beneficial to the species. Abundant evidence of beneficial mutations is all around us, if we consider the numbers of bacteria, viruses, and insects that are resistant to chemicals designed to kill them (beneficial, that is, to the pest, not to the humans who wish them dead or rendered harmless). Also, as we shall see very soon, mutational change conferring improvement is relative to all the external environmental components of a population and not only to the narrow focus of interaction between one life form and another, as in the case of antibiotic-resistant bacteria and their human hosts.

Recombination

While mutations produce new genetic information, recombination leads to new combinations or arrangements of existing alleles of the genetic complement (see Fig. 5.5). Although recombination has been known from the beginnings of genetic studies, it was not appreciated as a mechanism for increasing populational diversity until more recent years. Today we consider recombination to be of paramount importance in increasing variability and in introducing genotypes of exceedingly diverse nature into the genetic structure of populations. Indeed, it is impossible to conceive of evolutionary improvement at rates required for meaningful change in life forms without the intercession of recombinational opportunities. The discovery of mechanisms leading to gene recombination in the asexual prokaryotes and viruses has permitted us to understand better their evolutionary success and their ability to change rapidly and significantly in changing environments and under varied selection pressures. Unlike the slower introduction of variation by mutations, new genotypes may be introduced and be spread very rapidly in populations by recombination. The phenomenon leads to numerous new gene combinations derived from different cell lineages and may result in the insertion of one allele or of any number of alleles of the parental lines into a new genetic background.

Let us consider three different populations of a bacterial species, each of which carries a beneficial allele for one gene, which confers resistance or sensitivity to an antibiotic on the populations. Population number 1 is penicillin-resistant (pen^r) but sensitive to streptomycin (str^s) and to tetracycline (tet^s), and thus has the genotype pen^r str^s tet^s. For our purposes let us designate population number 2 to be resistant only to streptomycin (therefor pen^s str^r tet^s) and population number 3 to be tetracycline-resistant but sensitive to the other two drugs (therefore pen^s str^s tet^r). So long as each population encounters only the one antibiotic to which it is resistant, the group will flourish and transmit its inherited benefits to all descendant cells. Each of these mutations arose at a constant rate in the population, perhaps 1 per 100 million (1×10^{-8}), and became established in each population if there was a selective advantage such as the presence of the drug in the environment in which the population lives.

For any population resistant to all three antibiotics, owing to mutational changes, the chance of the occurrence of all three independent mutational events in the same cells would be derived from the product of the separate probabilities, or $10^{-8} \times 10^{-8} \times 10^{-8}$, or 1×10^{-24}. This is a vanishingly small chance of occurrence, and we would expect few or no populations to harbor such mutants, as a general rule, if this genotype is the result of mutations alone. This is the same as saying that we would expect to find very few occasions when 100 tosses of a penny led to its turning up heads every single time in 100 tries. The probability for the coin's landing in the heads position in 1 try is $\frac{1}{2}$. The chance of heads each time in 10 tosses is $\frac{1}{2}$ raised to the tenth power or $(\frac{1}{2})^{10}$, and its chance each time in 100 separate tosses would be $(\frac{1}{2})^{100}$. The general rule is that the probability that the separate independent events will occur simultaneously or concurrently is the product of the separate probabilities, or if a single probability equals x and the number of independent events equals n, then x^n equals the combined probability of these particular events occurring. There is a statistical chance for all possible combinations of events, but less probability for repeated events to occur relative to other alternatives in the spectrum of possibilities. In the example of the drug-resistant bacteria, the same principle prevails as with the penny tosses, so that $x = 10^{-8}$ and $n = 3$ independent events, or there are $(10^{-8})^3$ chances for a specific triple mutant to be present in any one of the populations we were discussing.

If alleles for resistance to one antibiotic are transferred to cells in another population which are resistant to a second antibiotic, then a new genotype will arise very quickly and spread through the population as descendant cells are produced bearing these same beneficial inherited traits. Furthermore, various recombinations may occur as different cells exchange genes, and many different genotypes may be produced, including some of great benefit to the population living in a particular environment. By such means it is highly probable that bacterial populations will become resistant to any number of antibiotics in a relatively short time and continue to flourish even in a changed or changing environment. In our original example, bacteria with the genotype *pen^r str^r tet^r* would arise with higher frequency as a function of gene recombination than by mutational events in the same cell. These phenomena are responsible in large measure for the very rapid change and spread of increasingly resistant bacterial strains in the 30 years since antibiotic therapy became common medical practice. In fact, the pattern of prescribing antibiotics as preventives, as well as curatives, has probably contributed the most to the ever-increasing problem of finding new antibiotics faster than variation and selection can act in bacterial strains to vitiate the therapy. This is not to say that antibiotic treatment should be stopped, but merely to introduce a proper note of caution against the indiscriminate medical practice of prescribing such drugs on more occasions than are warranted.

The evolutionary considerations are much more complex when we count the thousands and millions of beneficial mutations that must have occurred, rarely and at random, in different populations across the billions of years of biological history on Earth. The more infrequent incidents of recombination in asexual species compared with sexual forms may explain in part the slower rates of evolution during the first few billion years and the explosive increase in biological diversity during the past 600 million years. The intervention of sexual reproduction provided a regular mechanism for gene recombination and essentially ensured the production of genetic diversity in every generation in populations capable of such activity. Recombination is a concomitant of sexual reproduction because each new generation is the product of the fusion of nuclei from two different cells.

Reassortment of genes occurs during sexual reproduction as a consequence of two crucial processes: meiosis and fertilization (Fig. 6.4). Each sexually reproducing species undergoes the process of meiosis, during which the chromosome number is reduced by one-half in the resulting cells (usually gametes or spores) and different combinations of parental alleles may be distributed in one gene set to each meiotic cell product. When nuclei of two competent cells, such as gametes, fuse during a fertilization event, various genotypes are produced from random pairings of genetically diverse sex cells. It is this series of events that leads us to analyze the inheritance patterns in specific ways indicative of independent assortment, gene linkage, or such special transmission patterns as those with sex-linked

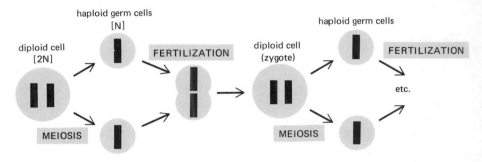

Figure 6.4   Distribution of chromosomes (and their genes) at meiosis and at fertilization in a sexual cycle underlie gene assortment and recombination and explain the constancy of the chromosome number in each generation of the species.

genes. In each case the pattern of transmission coincides wtih the chromosomal basis for the distribution of alleles; therefore we can interpret breeding data quite clearly in most crosses.

In addition to reassortment of genes in sexually reproducing species (which contain at least two different, nonhomologous, chromosomes) simply by random distributions of these chromosomes at meiosis and random recombinations at fertilization or nuclear fusion, recombinations may be produced by crossing-over between homologous chromosome segments (Fig. 6.5). Independent assortment does not occur in prokaryotes because they have only one linkage group. But both prokaryotes and eukaryotes, as well as the viruses, may experience recombination of linked genes because the crossing-over process is similar or identical in all these life forms. Since chromosomes are inherited as intact structures, the rearrangement of genes must involve some mechanism that reshuffles the contents of homologous chromosomes. The evidence favors a process of breakage and subsequent rejoining of broken chromosome ends, with no loss of genetic material under ordinary conditions. Depending on the location of a crossover and the particular alleles on the homologous chromosomes, new genotypes will be generated in a continuous spectrum of variation. If there are few crossovers then fewer new genotypes will be produced than if many crossover events occurred in a particular chromosome pair. Obviously the genetic expression of crossing-over occurrences depends on the degree of heterozygosity of the chromosomes. There may be crossing-over between chromosomes carrying identical alleles, and the products of such events would be genetically identical to each other and to the original parent chromosomes.

The inheritance pattern for linked genes is detectable by the general observation that loci on the same chromosome reassort less often than is the case for gene loci on different chromosomes. The usual approach in

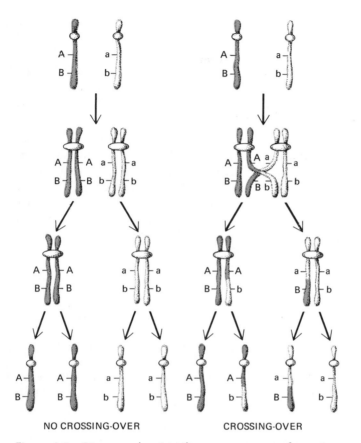

NO CROSSING-OVER   CROSSING-OVER

Figure 6.5   Diagram showing the consequences of crossing-over during meiosis versus no crossing-over event. Only one pair of homologous chromosomes is shown and only two of the many genes that occur on these chromosomes. In this case two kinds of gametes are produced when there has been no crossover but four kinds subsequent to a chromosome exchange of the sort depicted. The chromosome is shown as a single unit in its unduplicated stage and as a double structure when duplicated but held together by a common centromere.

determining whether genes are linked or not is to perform a backcross in diploid species or to interbreed suitably marked haploid populations. In either case one would find that parental classes and recombinant classes occurred in equal frequencies when genes were not linked, whereas parental classes would outnumber recombinants if the genes were linked. The percentage of recombinants derived from breeding analyses is translated directly into a measure of the distance between the linked genes. Thus, 1 percent recombination for two genes equals 1 map unit, 20 percent recombination means the two genes are 20 map units apart on the chromosome, and so forth. The recombinant frequency never exceeds 50 percent in a dihybrid cross (two genes differ in the parents), and this is the same percentage obtained for independently assorting genes. Many crosses involving as many genes as can be studied are needed to construct linkage maps of the chromosomes of the complement. More detailed discussions can be found in any genetics textbook. The main point for our consideration is that diversity may be generated by recombinations of linked genes through crossing-over, or by reassortment of genes on different chromosomes. The first mechanism may occur in either sexual or asexual species, in prokaryotes, eukaryotes, or viruses; the mechanism of independent assortment occurs typically in sexual eukaryotes and provides them with an added dimension for release of potential genetic diversity in populations. Together with mutation, these phenomena account for the principal means by which biological variation is maintained and modulated in populations.

Changes
in Chromosomes

We will consider three general categories of chromosome changes: (1) modifications leading to new chromosome numbers, either of whole sets of chromosomes or of individual chromosomes of a genome (chromosome set), as in polyploids and aneuploids, respectively; (2) changes in one or more chromosomes of the genome which have resulted from such gene rearrangements as occur in deletions, duplications, inversions, and translocations; and (3) alterations in chromosome numbers subsequent to breakage and rejoining sequences such as those involved in centric fusions. Many populations of plants and animals have been investigated using cytological and cytogenetic methods, and a substantial body of information is available.

POLYPLOIDY AND ANEUPLOIDY

It has been estimated that about 50 percent of the flowering plants are polyploid species. It is not at all unusual to find a series of ploidy levels in the species of some genus of angiosperms, and many of the cultivated crop plants, such as cotton, tobacco, potato, wheat, and others, exist as diploid, tetraploid, and sometimes hexaploid or higher polyploid populations. Odd numbers of chromosome sets also are found, but most triploid, pentaploid, or similar types are highly infertile or totally sterile species because of irregularities in chromosome distributions during meiosis. Aberrant chromosome complements either render the gametes inviable or the fertilization product incapable of developing to adulthood in virtually all polyploids with an odd number of genomes. Polyploidy is lacking or rare in natural populations of most sexually reproducing animal species, but there are some known polyploid groups among invertebrates. It generally is assumed that polyploid animals would be at a considerable disadvantage because of ensuing disturbances in the sex chromosome numbers and in the balance between autosome (nonsex chromosome) sets and kinds and numbers of sex chromosomes, and because of hormonal irregularities which would interfere significantly with differentiation and development of the organisms. Since most plants have no known sex chromosomes and since their hormonal regulation is independent of sexuality, fewer developmental problems would be encountered by polyploids in most of the groups of complex plants.

Aneuploidy occurs much less often than polyploidy, among both plants and animals. By aneuploidy we specifically mean an increase or decrease in one or more chromosomes of a genome, with no significant internal modification of the chromosome itself. Humans having 45 chromosomes instead of 46 thus would be monosomic for one chromosome of the complement, the other 22 pairs occurring as expected. In some cases there are 45 chromosomes for reasons other than loss of one linkage group, but generally one entire chromosome would be missing in monosomic individuals of any species. Trisomic individuals among humans would have 47 chromosomes in every somatic cell, one of the 23 kinds of chromosomes occurring in triplicate while the remaining 22 pairs would be normal (Fig. 6.6). A frequent trisomic type in humans is the individual with Down's syndrome, also known as a mongoloid. Such people have various developmental defects due to the extra chromosome. It is usually the case that addition or loss of a chromosome leads to lowered fitness of the individual and often to premature death or to reproductive failure, whether in humans or in other species. There is a very unusual situation in one of the flowering plant species, *Claytonia virginica* or spring beauty, in that chromosome numbers between 12 and about 190 have been found in members of the same species. Sometimes many different chromosome numbers may be found in the same population of these plants in open woodland areas. Fertility seems to be uncorrelated with variant chromosome numbers in *Claytonia,* which also is quite unusual.

Most species exist in populations that have the same chromosome number, although occasional variants arise. Generally there is no evolutionary advantage to aneuploidy of this classical sort, and we may consider it a less significant means for producing diversity in populations. Polyploidy, on the other hand, is a remarkably successful life style for many plant species and should be viewed as a significant device for modulating the level of diversity during the evolution of a species group or family of plants. An increase in the amount of genetic material should introduce a theoretically higher diversity potential, especially since we would expect originally identical

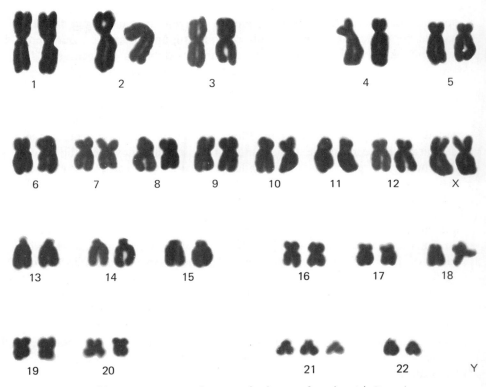

Figure 6.6   Chromosome complement of a human female with Down's syndrome. The chromosome number 21 is present in triplicate rather than as the usual pair in this diploid set of chromosomes. Although chromosome-21 and chromosome-22 are very similar in appearance, there is no doubt that the extra chromosome is number 21. (Photograph courtesy of L. J. Sciorra.)

chromosomes to undergo different mutations and ultimately come to direct different sets of life functions. This provides possible long-term advantages for increased variation, but there would be lower variability in polyploid populations early in their evolutionary history. Because there would be one allele of each gene on each of the homologous chromosomes of the genomes present, many fewer recessive combinations would be produced and there would be a slower release of expressed variation. In diploids it takes only two recessive alleles, one on each homologous chromosome of a pair, for phenotype expression. In a tetraploid four recessive alleles would have to be present in the same individual for the recessive phenotype to be expressed. Statistically fewer progeny would be produced with the recessive characteristic, thus limiting the diversity which is released on a short-term basis in the population.

Of course there would be situations in which limited expression of diversity would be advantageous; perhaps polyploidy became a significant evolutionary factor at particular times during the history of a species group. Among other possibilities, it may be that polyploid populations would have enjoyed selective advantages during the glacial and interglacial periods of the Pleistocene epoch, when fluctuating environments were prevalent. Lowered diversity usually is advantageous in changing environments, whereas higher levels of variability tend to confer advantages on populations living in relatively stable situations. It is a documented fact that many of the polyploid species of flowering plant groups occupy areas that once were glaciated, while their older relative species now are found in regions that were not covered by glaciers during the Pleistocene. Such a correlation can be explained in various ways, but one aspect may be related to the storage and release of variability in short-term and long-term perspectives of evolutionary success in polyploid species groups.

The origin of polyploid or aneuploid individuals probably is traceable to events at meiosis. If gametes are produced during an aberrant meiosis, they may contain one chromosome more or less than a complete genome, and on fusion with another gamete, whether normal or aberrant, they may produce an individual with a monosomic or trisomic chromosome content. Similarly, gametes or other cells competent to fuse and produce a zygote may contain an unreduced number of chromosomes because of some failure of meiosis. If two gametes fuse, a tetraploid may be be produced if the two gametes are diploid, or another level of polyploidy may result, depending on the number of chromosome sets in the cells that combine to yield individuals of the next generation.

There are various ways in which polyploidy can arise, and there is a large body of experimental and observational literature on this subject. For many kinds of cultivated plants it has been possible to create polyploids through the use of chemicals such as colchicine, which is an alkaloid derivative of the autumn crocus and a common medication for the treatment of gout in people. Flower and fruit size tend to be larger in polyploids and enhance the usefulness or desirability of the plants. Sterile polyploids may produce flowers for long periods of time because no fruit or seeds are produced and growth and development therefore continue to be channeled to flowers. Many of the prize orchids are sterile triploid or pentaploid forms, not only producing large or numerous flowers but also capable of producing flowers for long intervals of time in the life of the perennial plant. Among the crop plants it is usually the tetraploid or hexaploid species that is cultivated in modern times; the wild species, if still extant, usually are diploid. Greater yield almost certainly accrued with the appearance of a polyploid variant, and such polyploids apparently were selected and used in place of the ancestral forms. Not all crop plants, by any means, are polyploid; species such as corn or tomatoes are diploid, as are a number of other important agricultural plants.

CHROMOSOME ABERRATIONS

Of the four general types of aberration, those long considered to be the least significant in evolution were *deletions* and *duplications*. There was very little evidence for these changes in natural populations because of the technical difficulties in detecting either of them. On theoretical grounds it was considered that loss of genes by deletion or gain of genes by duplication would be unlikely to lead to significant improvement in populations, at least in terms of short-term advantages. Since recent studies have emphasized the prevalence of gene duplications, as we discussed earlier in Chapter 5, the evidence should be reviewed from this new perspective. Duplications probably are a source of new diversity, since different mutations will occur and eventually lead to added information content modified from the supplemented genetic material. In addition to providing raw materials which may increase diversity, repetitive DNA occurs commonly in eukaryote genomes. If the genetic redundancy is a function of chromosome duplications, then we should consider this parameter of evolutionary change and perhaps revise the older conclusions, which were based on limited observations. In the sense that duplications may lead to greater potential for producing diversity in long-term evolution, this mechanism of chromosome change really should be evaluated separately from other types of aberrations.

Inversions and *translocations* have been considered to be the most important factors in modifying the level of population diversity. Cytological evidence for these aberrations is easier to obtain and interpret; thus much of the literature centers around these forms of chromosome change in a variety of plant and animal species. An inversion or a reciprocal translocation usually has little effect on the phenotypic expression of the genotype because all or most of the gene content still is present, though rearranged. Inversions involve a displacement by 180° of a segment of one chromosome so that two breaks must have occurred and been rejoined, establishing new linkage relations (Fig. 6.7). It is possible to detect inversions by genetic analysis, since particular genes in or near the inverted region will show altered linkage distances (determined by percentage recombinants in the progeny). Cytological evidence has come principally from studies of natural

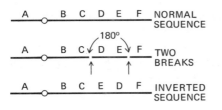

Figure 6.7   Diagram representing the formation of a chromosome inversion. The letters represent an arbitrary sequence of gene regions, and the open circle connecting the two arms of the chromosome depicts the region of the chromosome (centromere or kinetochore) which functions in movement to the poles during nuclear division.

populations of various species of *Drosophila*. The huge polytene chromosomes of the salivary glands of the larvae permit detailed analysis of genetic sequences from comparisons of the banding patterns that occur. The banding is consistent and can be used to construct chromosome maps that coincide with maps generated from breeding analysis and recombinant frequencies for linked genes. Therefore, modifications in banding patterns in the salivary gland chromosome can be used directly to evaluate inversion phenomena.

One consequence of inversions in populations is that inversion heterozygotes—that is, individuals with one inverted and one noninverted chromosome for a pair of homologues—produce progenies with reduced variability. This is due in part to the reduction in crossing-over frequency in inverted regions of the synapsed chromosomes during meiosis, thus producing fewer recombinations of linked genes on the affected chromosomes. Also, some kinds of crossing-over events involving the aberrant chromosome in heterozygotes lead to defective gene complements in the gametes or spores produced during meiosis. Such cells would be inviable, and since these are precisely the potential recombinants, a further reduction in genotype diversity is produced in populations that contain inversion heterozygotes. Inversions therefore act to reduce population diversity. Under some conditions lowered diversity would be advantageous to the immediate requirements for success of a population. We shall discuss this further in Chapter 9.

Translocations may be simple or reciprocal, with the latter being much more common in most populations. In a simple translocation a piece of one chromosome becomes a part of a nonhomologous chromosome. Reciprocal translocations involve breaks in two nonhomologous chromosomes also, but there is an exchange of parts so that two new linkage groups are produced from the rearranged materials of the original chromosomes (Fig. 6.8). Little genetic material is lost, as a rule, and there usually is no reduction in chromosome number for the population. Reciprocal translocations may involve only two chromosomes or, in an unusual genus such as *Oenothera*, all the chromosomes of the genome may have undergone exchanges. (One of De Vries's problems in interpreting his genetic data was that he was unaware of the eccentricity of inheritance patterns that resulted from the presence of reciprocally translocated chromosomes in *Oenothera*. In 1900 it was not known that inheritance patterns were chromosomal in basis.) As with inversions, the main concern is with translocation heterozygotes and not with homozygous normal or aberrant populations. Analysis of chromosome pairing during meiosis indicates very clearly whether or not translocation heterozygosity is present. The translocated and nontranslocated chromosomes undergo synapsis so that homologous chromosome regions pair, and a complex chromosome configuration can be seen under the microscope. Depending on the numbers of chromosomes

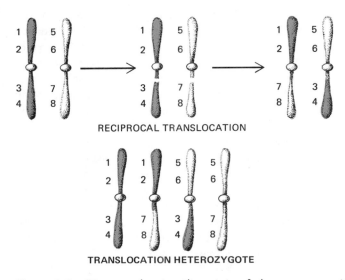

RECIPROCAL TRANSLOCATION

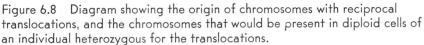

TRANSLOCATION HETEROZYGOTE

Figure 6.8   Diagram showing the origin of chromosomes with reciprocal translocations, and the chromosomes that would be present in diploid cells of an individual heterozygous for the translocations.

that have exchanged segments, one may find chromosomes joined in configurations of various numbers. In *Oenothera* species all 7 chromosomes of the genome are translocation types so that all 14 chromosomes in heterozygotes are joined together in a single complex of chromosomes during the first division of meiosis. Circles and chains of chromosomes often can be seen in microscope slide preparations; the number involved in the configuration provides the best clue to the extent of translocation exchange that has occurred.

As was the case for inversion heterozygotes, translocation heterozygotes produce less-diverse progeny because of meiotic irregularities and the inviability of gametes or spores with deficient or duplicated genetic material. The selective advantages of lower diversity can be explained, as we shall see later. That there is a selective advantage can be discerned in some cases because of characteristics of the population which enforce a high proportion of heterozygotes to be produced, thus further ensuring limited release of diversity. In *Oenothera* for example, there is self-fertilization, so one would expect progenies to contain structural homozygotes as well as heterozygotes after the segregation and recombination of chromosomes at meiosis and fertilization. But there is a system of balanced lethal alleles, which operates in such a way that homozygotes are inviable and the only surviving progeny are the same translocation heterozygotes generation after generation. This mechanism enforces a lower level of variability on *Oenothera* populations.

Although this group of species represents an extreme case, the principle can be extended to other examples of translocations in which limited release of diversity is advantageous for the particular life style in which the populations operate best.

CENTRIC FUSIONS AND REDUCTION IN CHROMOSOME NUMBER

Comparative analysis of groups of related species often reveals an apparent aneuploid series in the genus. Species of *Drosophila,* for example, may have three, four, five, or six pairs of chromosomes (Fig. 6.9). Careful study of the size and morphology of each chromosome in these genomes reveals that aneuploidy is not responsible for the variation in chromosome number; that is, there has not been a gain or loss of whole chromosomes of a complement. There appears instead to have been a series of rearrangements of the genetic material such that originally separate chromosomes have evolved toward fewer and larger ones which contain most of the genome information. In *Drosophila* it is possible to be more certain of the rearrangements because banding patterns of the larval salivary gland chromosomes permit fairly accurate comparisons. The evolutionary changes apparently have been due to reciprocal breaks and subsequent rejoinings of nonhomologous chromosome segments, very much like those we discussed for translocations. In the present situation, however, there often is such substantial transfer that one new chromosome comes to contain most of the genetic material that was present originally in two chromosomes. The remaining chromosome fragments may either be lost or remain in the complement as a very small structure.

A frequent event in these rearrangements is that two chromosomes, each with the centromere (kinetochore) very near the end of the structure,

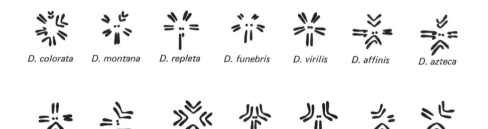

D. colorata   D. montana   D. repleta   D. funebris   D. virilis   D. affinis   D. azteca

D. pseudoobscura   D. americana   D. ananassae   D. melanogaster   D. putrida   D. prosaltans   D. willistoni

Figure 6.9   Chromosome complements of 14 different species of *Drosophila,* showing the aneuploid series varying from three to six pairs of chromosomes in the diploid cells. These genomes are all from male insects, showing the X and Y chromosomes at the bottom of each drawing.

undergo a change leading to a much larger chromosome with a median centromere. Centric fusion has been the general term applied to this particular kind of chromosome reorganization phenomenon (Fig. 6.10). In fact, centric fusions are so common in animal species that it has come to be stipulated that chromosomes that are acrocentric (centromere near one end of the chromosome) are more "primitive" while metacentrics (median centromere location) represent a more "advanced" trait, since they are derived from acrocentrics. If this concept is stretched too far we would have to consider humans as less advanced than chimpanzees because these apes have no acrocentric chromosomes whereas we have five (chromosomes 13, 14, 15, 21, and 22), plus the acrocentric Y chromosome in males (Fig. 6.11).

There is no doubt that centric fusions are significant events in the evolution of the genetic system in animals, but one must be cautious in basing a far-reaching conclusion on only one line of evidence. According to G. Ledyard Stebbins, who has contributed substantially to our knowledge and understanding of chromosome evolution in plants, centric fusions have been important in the evolution of some plant species, but they probably are a less significant factor in plants than in animal species. For either plant or animal species, however, there is a common feature of evolutionary consequence in the rearrangement of the genetic material into fewer linkage groups. Phrasing the matter in this fashion should make it obvious that recombination potential is reduced when the chromosome number is reduced. There would be fewer recombinants from independent assortment of genes on different chromosomes and a greater reliance on recombination through crossing-over between linked genes in a genome of reduced size. Once again, there would be lowered release of genetic diversity in successive generations and a greater probability that parental gene combinations would predominate in these populations. Under some conditions there would be high selective advantage for such a genetic system, in which successful genotypes predominate and new and untested genotypes are kept to a minimum in a species, generation after generation.

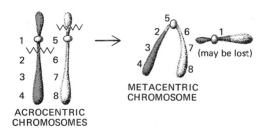

ACROCENTRIC
CHROMOSOMES

METACENTRIC
CHROMOSOME

(may be lost)

Figure 6.10 Schematic diagram showing the origin of one large chromosome from two nonhomologous chromosomes by the process of centric fusion. The small chromosome that results may be retained in the genome, but more often it is lost if its gene content is not essential.

Storage
of Diversity

As we have discussed already, mutations are rare events, which occur at random and usually are harmful to the population when they are expressed. Still, mutation is the only known mechanism by which new genetic information is produced, so it is the primary source of diversity in species evolution. There are two major problems that must be solved: (1) retention of new variability produced by mutation, since this is the raw material for evolutionary progress, and (2) protection of the population from immediate expression of harmful alleles that arise by mutation. Both problems can be solved by a single mechanism, namely, the development of diploidy and the extension of the time during the life cycle when the diploid condition prevails. Since the wild type allele usually acts as a dominant, and since most mutant alleles act as recessive to the wild type alternatives in a diploid organism, new mutations would be masked from expression.

Diploidy is a regular feature of sexually reproducing species because gamete fusion leads to the fusion of separate nuclei and the constitution of a single zygote nucleus containing the genomes of both parent lines. The longer the interval of time between the production of the diploid zygote and the process of meiosis in which the chromosome number is restored to the haploid level, the longer the diploid state will prevail during the life cycle. In some protists, especially among the protozoa, and in the more highly evolved animal and plant groups, the principal condition during the life cycle is diploidy. In vertebrates only the gamete itself is haploid; fusion occurs almost immediately after gamete production at meiosis, and the adult organism is diploid throughout its life span. Among the green plants there is an evolutionary progression in which the diploid state occupies an increasing proportion of the life cycle, the haploid stage being reduced concomitantly. The mosses primarily are haploid plants and the diploid stage is transiently associated with the persistent haploid individual. Among the ferns the familiar plant is diploid and predominant, but there is a haploid stage in which another plant form, physically separate from the diploid, is produced. This haploid gametophyte is very important because it is the gamete-producing stage, and it is susceptible to hazards because it is dependent on water; it is, moreover, a free-living individual. Further reduction in the size and life-span of the haploid stage has occurred in the gymnosperms so that this stage is microscopic in size and is nurtured by the predominant diploid plant, such as a pine tree. Among the angiosperms the haploid has been reduced to just a few cells in size and its development takes place entirely within the diploid flowering plant.

Species that are predominantly haploid also have a mechanism for storing genic diversity even though the nucleus contains only one genome. Many of

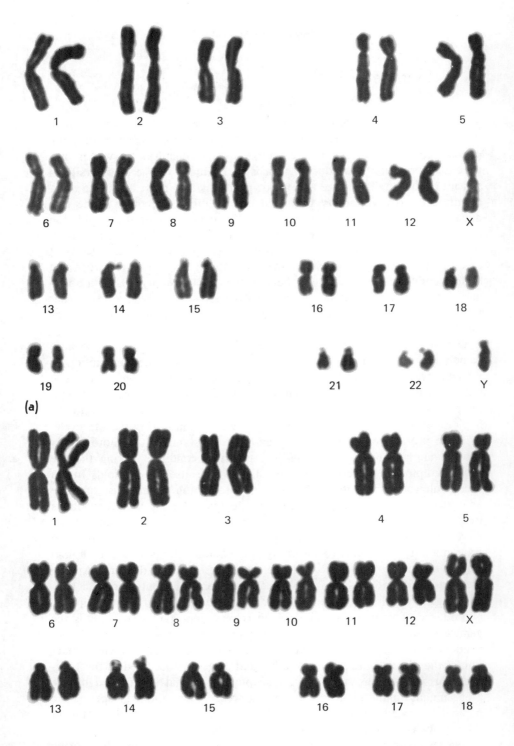

the fungi, for example, are multinucleate organisms. Since it is entirely possible for genetically different nuclei to occupy the same cell, the genetic constitution of a multinucleate organism may be such that recessive alleles in some nuclei can be masked from expression by the dominant alleles that may occur in other nuclei. Individuals carrying nuclei with different alleles for the same gene complement are called heterokaryons. Homokaryons have identical nuclei with the same alleles present in each genome present in each nucleus of the individual. There is a potential for diversity via heterokaryon formation and there also is the possibility for fusions of allelically different nuclei. Nuclear fusions provide opportunities for recombinational events upon subsequent haploidization and the segregation of haploid gene sets into new nuclei. Recombination may occur as a function of independent assortment or of crossing-over, both of which increase diversity. The frequency of crossing-over in such somatic cells is considerably lower than one finds for meiotic cells, but it has been reported to occur in as many as 2 percent of the mitotic nuclear divisions in some species of multinucleate, multicellular fungi. In fact, crossing-over during mitosis has permitted the construction of linkage maps for at least one such species, the common airborne contaminant *Aspergillus nidulans*.

The only mutations of evolutionary significance are those transmitted through the germ line to succeeding generations. Mutations may occur in any cell at all, but somatic mutant cells are of consequence only in the immediate individual in which they appear; there is no transmission of such diversity to descendant generations. The importance of the gametes, the only connecting link between sexual generations from a genetic standpoint, is emphasized in many examples of evolutionary development. The organs in which the gametes are produced, such as ovary and testis in many animals and analogous structures in plants, often are protected by external layers of tissues. In amphibians the eggs and sperm are shed from the adults, and fertilization takes place externally. During evolution continual improvement has occurred among more advanced forms. Among the higher vertebrates the egg is retained within the female parent, it is fertilized internally, and the development of the fertilized egg proceeds within the body of the female parent. We will consider such evolutionary pathways in another chapter.

This chapter concludes with an apparent paradox. Mutation is rare, ran-

---

Figure 6.11   Chromosome complement of (a) a human male (XY) and (b) a human female (XX) showing the varying position of the centromere in different members of the complements. Chromosomes 13, 14, 15, 21, and 22 are acrocentric (as is the Y); metacentrics are represented by chromosomes 1, 2, 3, and others, whereas a submetacentric chromosome resembles numbers 4 and 5. (Photographs courtesy of L. J. Sciorra.)

dom, and usually detrimental, yet the phenomenon produces the only new genetic materials which lead to evolutionary improvement in all life forms. Improvement has been the persistent theme in evolution, but it seems entirely improbable in view of the nature of mutation. How can we reconcile these opposing processes to solve the paradox? Lamarck has been proven incorrect in his proposal that only beneficial inherited changes have occurred, in response to the needs of the species. We will consider next the concept of natural selection, essentially as proposed by Charles Darwin and Alfred Russell Wallace in 1859, and we shall present the evidence in support of the action of selection on diversity leading to adaptation during the evolution of life from its beginnings on this planet to the present day.

Some Suggested Readings

Benzer, S., "The fine structure of the gene." *Scientific American* 206 (January, 1962), 70.

DeRobertis, E. D. P., W. W. Nowinski, and F. A. Saez, *Cell Biology.* 5th ed. Philadelphia, Saunders, 1970, pp. 255–327.

Ingram, V. I., "How do genes act?" *Scientific American* 198 (January, 1958), 68.

Stebbins, G. L., *Processes of Organic Evolution.* 2nd ed. Englewood Cliffs, New Jersey, Prentice-Hall, 1971, pp. 18–60.

Stebbins, G. L., *Chromosomal Evolution in Higher Plants.* Reading, Massachusetts, Addison-Wesley, 1971.

Yanofsky, C., "Gene structure and protein structure." *Scientific American* 216 (May, 1967), 80.

CHAPTER 7
NATURAL SELECTION

The paradox of evolutionary improvement based on rare and generally harmful mutations can be resolved by inserting the principle of selection as a guiding force that acts on population diversity in the direction of adaptation. Both Charles Darwin and Alfred Russell Wallace proposed essentially similar concepts, in each case based on documented observations of a wide variety. The two British scholars entered into communication that led to their joint appearance at a meeting of the Royal Society in London to report on their studies. We ascribe the ideas almost entirely to Darwin, as did his contemporaries, because of the voluminous publication in 1859 and the subsequent articles and books he wrote on the subject. His book caused a vigorous controversy between proponents of Natural Selection, such as Sir Thomas Huxley, and the clerics, naturalists, geologists, and others who rejected the theory on one or more criteria. The theological arguments were the most prominent, since the concept of Special Creation was widely accepted at that time. In fact a considerable direction to Darwin's thinking was provided by the difficulties of resolving the geological, paleontological, and biological data with the doctrine of Special Creation.

To the public it was repugnant to consider the inclusion of other species in their ancestry, especially such "brutes" as the apes. This particular thorn continued to bother people of various persuasions. The premise of our animal ancestry was not welcomed into church doctrines even to very recent times and is not accepted by some denominations today. Many sticky questions delayed the acceptance of Darwinian ideas by the biologists and naturalists of his time. How could wild species be so diverse when breeding studies had shown that the progenies of inherently different parents usually were less variable than the parents themselves, as when black and white animals produced gray offspring? There appeared to be a loss of variability on interbreeding (as in the substitution of grays for the black and the white parent groups), and this puzzle was not solved until many years later when the new discipline of genetics provided the answers. How could gradual changes have led to new species when the known species were so well defined and distinct from one another, and appeared to have been distinct for a very long time? Of course, nineteenth-century questions have been answered for the most part by twentieth-century observations and experiments. But different series of questions arose at intervals after 1859, and a number of questions are still posed today in relation to some aspects of the theory of natural selection. We do not reject the entire theory today just because we have not been able to understand some particular evolutionary phenomenon in terms of Darwinian selection. But we do search for new dimensions to the principle of selection and to a refinement of those concepts that have not proven to be sufficiently applicable to the interpretation of some difficult biological problem.

After the rediscovery of Mendel's work in 1900 by Correns, Tschermak, and De Vries, a new dimension of difficulty arose in accepting the principles of natural selection. The mutations studied in the laboratory appeared rarely and usually were harmful, as well as seeming to affect relatively few significant properties of the organisms analyzed. Most of the natural variation that could be observed and studied was of the *continuous* type, that is, a continuing graded spectrum of variation such as occurs for height or weight or many other multigenic traits. In the laboratory it was *discontinuous* variation that was observed, discrete classes of phenotypes with little or no intermediate quality, such as red eyes or white eyes in *Drosophila* or green versus yellow seeds in the garden pea. There seemed to be a different quality of variation in laboratory populations as compared with wild species. In the 1930s and afterward methods were developed to explain the rates of evolutionary change and the patterns of genetic variety in populations. Biometrics was born and the works of J. B. S. Haldane, Sewall Wright, Sir Ronald Fisher, and others phrased the problems in mathematical terms and for populations rather than for individual organisms. Population genetics has continued to develop and to provide the biometric perspective and the mathematical methods that can be applied to evolutionary questions and problems. In the 1940s another dimension was added to studies of evolutionary phenomena and to speciation in particular. The concept was developed of isolation of breeding populations as a prerequisite to the fixation of species variability and ultimate discontinuity between species. This premise proved so significant to the understanding of species origin and development that we now consider the modern concepts of evolution to be neo-Darwinian, that is, Darwin's proposals plus others added during the past hundred years. These concepts comprise a substantial and coherent set of principles by which most evolutionary phenomena may be understood. We will consider some of these ideas in subsequent chapters.

Basic Principles
of Selection

The phrase "survival of the fittest" really is a tautology and thus not explanatory. Survival means the perpetuation of a line and therefore it means leaving the most individuals to perpetuate the line. According to our

modern concepts, the fittest are those that leave the largest number of offspring. Taken together, natural selection when viewed as survival of the fittest means that the populations that leave the most offspring are those that leave the most offspring. This is tautological, a repetition of words without further clarification.

A more meaningful concept of natural selection incorporates the importance of reproduction rather than survival, since reproduction leads to descendants and to the transmission of genetic properties, whereas survival may or may not lead to the production of offspring. An appropriate view or definition of natural selection would be: *differential reproduction of genetically diverse populations.* By this phrasing we imply that natural variation occurs in populations of organisms such that there exist different genotypes and different frequencies of the allelic forms of the genes that specify the characteristic traits of each population. The relative frequencies of alleles and genotypes will change with time because some members of the population leave more progeny than others. Those leaving the most progeny are most fit in the particular environment at the particular time we observe the populations. The fitness is a function of a greater suitability based on hereditary traits that render some individuals more likely to achieve the stage of reproduction than would others in a population.

The population is the basic unit of evolutionary change, since all the genes in their varied allelic forms are shared by all components of the group. The gene pool, that is, the combined genetic potential of the population and of all the populations that comprise a species, is the changing quality and quantity. Depending on differential reproduction, which is a function of genetic properties, populations undergo evolutionary changes such that improvement usually occurs. Selection acting on natural diversity the guiding principle for evolutionary improvement or adaptation. With selection, evolution could not have occurred in the ways we have obser Indeed, as stated succinctly by Sir Ronald Fisher, natural selection mechanism for generating an exceedingly high degree of improbab Given that mutations are rare and generally harmful, improvement occur only by means of differential reproduction. Detrimental alleles transmitted to few or no progeny, because such alleles render individuals less likely or unlikely to reproduce; beneficial alleles on the other hand, despite their rarity, are transmitted to descendants because they confer an advantage on the individuals and make reproduction a more likely occurrence. Selection modulates the frequency of alleles in the gene pool of the population, thus providing a directiveness to evolutionary change. Rates of change due to selection generally are greater than rates of mutation and therefore account for the particular pattern of change in all categories of life forms. When a rare beneficial allele appears, it can spread in the population at a fairly rapid rate because parents with this allele will leave more descendants (more copies of this allele); in time the beneficial allele may become more common in the gene pool than its less beneficial alternative

forms of the gene. Harmful mutant alleles would not tend to accumulate in the gene pool, because fewer descendants would be produced by those members harboring such genetic traits. The situation is more complex for diploid species, because recessive alleles would be masked from immediate expression by the dominant wild type forms. We would be more accurate if we acknowledged that detrimental alleles could accumulate, since they may not be subject to selection in most diploids, but that their frequency would be kept down by selective forces. This subject will be discussed at greater length in Chapter 8.

Two fundamental questions had to be answered if the evolutionary process of selection acting on natural diversity was to be accepted and applied: (1) Is there sufficient variation in nature to account for changes in populations during evolution? And (2) Is variation (mutation) truly random? Casual inspection of a species reveals some diversity, but not much. The obvious diversity is interspecific rather than intraspecific. Part of the difficulty almost certainly is due to our insensitivity to existing variability in populations of organisms with which we are unfamiliar; all giraffes look alike unless we come to know some of them as individuals. Furthermore, most variation in diploid species is hidden, since recessive alleles are not evident in heterozygotes as a general rule and are expressed only in homozygous genotypes.

A comprehensive series of studies using genetic methods to analyze natural variation was conducted during the 1920s and 1930s by Soviet geneticists, including Dubinin and Timofeef-Ressovsky, and by such American investigators as Theodosius Dobzhansky. Their data revealed a wealth of diversity in the samples from natural populations that were taken to the laboratory for study. In addition to the demonstration of many recessive ... s, cytological analysis of banding patterns in salivary gland chromo- ... s revealed many structural variations in various *Drosophila* species ... their component populations. These latter studies had to await the ... ification of the giant structures as chromosomes, first made known in ... and then utilized to construct linkage maps by Calvin Bridges and ... pioneer geneticists working with *Drosophila*.

... any lines of evidence showed conclusively that mutants that occurred in ... tural populations were of the same kinds as those that occurred in the laboratory, either spontaneously or after induction by chemical and physical agents. Genetic analysis of variation was extended substantially, since investigators could utilize variants from any source to study the nature of mutation and the gene. These pioneer studies showed that diversity within species was of the same sort as diversity between species, all together comprising pools of genetic variation that served as the raw material for evolutionary change. Modern molecular analysis of the gene and mutation has amplified these features of genetic material and strengthened our evidence of a common denominator of inherited diversity in all life forms and, therefore, a common basis for evolutionary processes in all species.

Since spontaneous variants and induced mutants were shown to be entirely comparable, experimental analysis could be initiated to determine whether or not genetic change was a random process. Extensive investigations of this question were conducted during the 1930s and 1940s especially, and all the evidence pointed to mutation as a random event lacking predictability in every system studied. Every measurable and observable trait of a population was susceptible to mutational alteration, which was further evidence of the randomness of the processes leading to inherited change. We can predict that mutations will occur, but not when or where they will take place or the direction of the change relative to benefit or harm for the population. The experimental design devised by H. J. Muller using *ClB* stocks of *Drosophila* (see Fig. 6.3) could be used to demonstrate the existence of natural variation and to determine whether spontaneous and induced mutations were comparable, and with what frequency the different categories of mutants appeared in populations. The method was used specifically to analyze genes on the X chromosome of *Drosophila,* and some question was raised as to whether sex-linked mutations could be considered typical of all mutations, including those on the commoner autosomes of species. Another experimental design was devised to explore the mutational phenomena on autosomes of *Drosophila*; in that case too there was no difference between spontaneous and induced mutations, laboratory and wild populations, or sex chromosomes and autosomes.

The experiments first reported in 1927 by H. J. Muller and in 1928 by L. J. Stadler showed conclusively that x rays led to an increased rate of mutation in multicellular organisms. Muller's studies using *Drosophila* and Stadler's using barley plants were comparable and provided the definitive evidence of increased frequency of mutations as a consequence of radiation. Equally significant were their studies demonstrating that the effects of radiation were cumulative, and that similar effects were produced on the genotype whether a particular total dose of radiation was presented over a short or a long interval of time or in larger or smaller individual amounts to achieve the same total exposure. The dangers of overexposure to radiation and other mutagenic agents have a fundamental long-term implication for descendant generations, as well as the more obvious effects on the exposed individuals. The clinical symptoms of radiation damage represent only one dimension of the harmful effects to organisms. The more serious implications, however, are in the increase in mutational load for a population. Since most mutations are detrimental to some degree, exposures to mutagenic agents should be considered as deleterious to prospects for continued improvement in such populations. The sinister threat of nuclear holocausts leading to death and destruction, like the horrors visited on Hiroshima and Nagasaki in 1945, is even more numbing from the long-term view. The damage to life forms as a consequence of increased genetic loads, in the form of increased levels of harmful recessive alleles in gene pools of species, would haunt the surviving populations for hundreds or thousands of generations afterward—assuming there would be survivors on the planet.

Lamarck Versus Darwin: Experimental Evidence

The evidence in support of selection acting on random natural variation appears to be overwhelming, but the concept is repeatedly brought into question. One often-cited example of evolutionary change involves the peppered moth species of Great Britain and Ireland. The moth occurs principally as a dark or a light phase variety, and has been collected for more than a hundred years by amateur and professional biologists. Such collections provide documented evidence of the relative frequencies of these two varieties over a long period of time in many parts of that region. There has been a significant increase in the frequency of the dark variety during the past hundred years in regions that have become increasingly industrialized and urbanized, and a consistently high proportion of the light moths in areas that have not experienced such changes in economy and human population patterns.

Based on principles of evolutionary processes, the explanation for the change in frequency of dark and light moths appears rather straightforward. Industrialized areas become increasingly sooty, and many resting places used by the moths, such as trees or buildings, have darkened over the years. The moth populations included both dark and light varieties all along, as can be determined from collections, but there would be greater selective advantage for the dark phase moth in areas with darker backgrounds than there would be for the light type. The darker moths would be less visible to their predators because of more effective camouflage against darker backgrounds, whereas the light moths would be more vulnerable in these locations. Conversely, the lighter moths would enjoy greater selective advantages on trees covered by light-colored lichens and thus would be less susceptible to predation than would the darker moths resting on a pale background. There would be an increase in frequency of darker moths in darker habitats because these would be more likely to escape predation, live to reproduce, and transmit the alleles for darker coloration to descendant generations. Lighter moths would become fewer because fewer would achieve reproduction age and so less of the light phase variety would be produced in each generation. Some of the lighter moths would persist in many areas because no habitat is so uniform as to lack some lighter resting places on which these insects might escape detection. There would be relatively little change in the proportion of light and dark varieties in rural or unindustrialized areas, thus providing a basis for explaining the predominance of the lighter moths in such locales.

This would appear to be a classic example of evolution in action, since the alternative allelic forms of the gene for coloration have changed in frequency in correspondence with altered environments and selection pressures. The alleles for pigmentation arose at random by mutation, and their frequencies, as well as genotype frequencies, would increase or decrease depending on the relative selective advantages in different environments, which vary in habitats, food supply, predators, and other factors. The species is perpetuated as a consequence of its genetic potential and variability as acted upon by selection. The dark form is more adaptive in some environments and the light form in others. The store of diversity in the populations of the species is an essential component for long-term success in changing environments (selection pressures).

Did the mutations really arise at random in these moths? Is it possible that the mutations leading to darker coloration appeared in response to the needs of some moth populations to survive the changing habitat? Can we really be sure that the specific mutations were undirected and unrelated to the needs of the species at the time they first occurred? Questions such as these are asked repeatedly, evidence that Lamarckian optimism persists. Various examples of genetically altered populations have been eyed suspiciously despite the wealth of evidence from the genetic analyses of 1900 to 1940, the great classical period of the discipline. This reluctance to accept the principles of selection acting on random variation does not emanate from acceptable experimental evidence in support of Lamarckian doctrines. The burden of proof generally is demanded of Darwinian evolutionists rather than from proponents of Lamarck, and modern anti-Lamarckians are required again and again to provide further substantiating evidence to bolster their interpretations.

The word "interpretation" causes the key difficulty. After all, each population studied is one in which the mutations already are present. How can we be sure that they arose completely at random and not in response to environmental stimuli for specific change? It is somewhat simpler to rebuke these skeptics now because of the strong basis of molecular genetic information, which underlies evolution in all known life forms. As a matter of historical interest, the most skeptical of all biologists were those who studied population dynamics of bacteria and other microorganisms. The phenomenal rapidity with which bacterial populations may change seemed to be in violation of the principles of gradual evolutionary change as a function of rare, random mutations. It was possible to demonstrate on repeated occasions that a culture of virus-sensitive or chemical-sensitive bacteria could change overnight, in the presence of these external agents, to become entirely resistant to further exposures to these same agents. This seemed so unlike selection of random variants and so much more akin to the Lamarckian doctrine of a specific beneficial change leading to improvement in the population and to the perpetuation of new inherent properties even when the original stimulus was removed at a later time. Was it possible that there was some truth to Lamarck's theory, which could be gleaned from bacterial studies but which was undetectable from genetic analyses of higher forms of life such as insects, mammals, or plants?

In 1943 Salvador Luria and Max Delbrück published the results of their experiments that were designed to answer just such questions as these. The design of the Fluctuation Test experiment (Fig. 7.1) can be explained most easily as a statistical analysis that permits the data to be interpreted

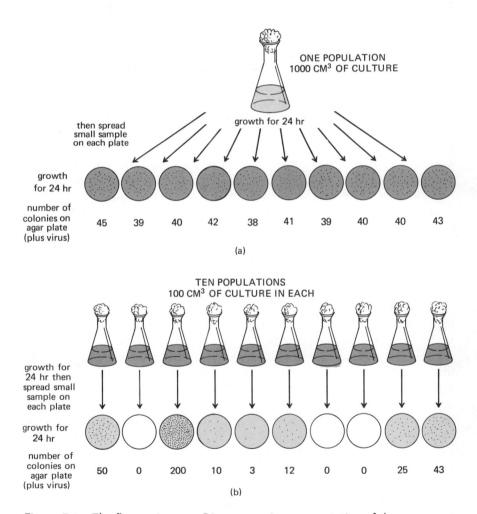

Figure 7.1   The fluctuation test. Diagrammatic representation of the experimental design of the Luria–Delbrück Fluctuation Test. Samples taken from one large population are compared with a comparable number taken from numerous small populations.

on only one of the two theories; either the data support Darwin or they support Lamarck, but not both. The basic question to be answered was whether beneficial change results from specific directed mutation (Lamarck) or from a process involving selection of preexisting spontaneous mutants in an altered environment (Darwin). The answer to this question presumably would be directly applicable to the bacteria under investigation and would be similarly applicable to any life form, since mutation is a fundamental property of all genetic material.

The experiment was arranged so that populations of the colon bacillus, *Escherichia coli,* were grown in culture under optimum growth conditions, except that viruses to which the bacteria are sensitive were included in the environment of these growing cells. Although the wild type *E. coli* dies in the presence of specific viruses, mutants were known that could flourish under these same conditions. It was mutants of this sort that were the object of the search, but the particular design of the experiment would make it possible to interpret the origin of the mutants, either as induced in the presence of the virus or as spontaneous variants that existed in the wild type populations whether or not viruses also were present.

One series of cultures was grown in large flasks containing nutrient medium, to which viruses were added at the beginning of the growth period. Another set of cultures was grown under identical conditions for the same 24-hour interval, except that the same total volume of cells in culture was distributed among a number of smaller flasks after inoculation rather than remaining in one flask. After 24 hours samples were removed from the cultures and plated on solid media so that each viable cell could develop in situ into a visible colony. Each colony of cells represented all the descendants of a single original cell removed from the culture, but only viable cells could reproduce and develop into colonies when put on solid growth medium, such as nutrient agar. Counts of colony numbers on the plates, after another 24 hours to permit their development subsequent to removal of the samples from the flasks of liquid medium, could be translated directly into the frequencies of virus-resistant cells present in the original flasks. The control cell cultures in these experiments were grown in flasks lacking viruses, so that comparisons could be made and information could be obtained on numbers of sensitive cells and on growth conditions during the experiment. Those cells grown in the presence of the viruses were sampled to determine the origin of the mutants recovered at the end of the experiment.

The rationale underlying the experimental design was as follows. If mutations were induced specifically in response to the virus, then we would expect to find the same frequency of virus-resistant mutants whether we took many samples from one large population or a single sample from each of many smaller populations of the same total volume of cultures grown in the presence of viruses. The same stimulus for change (the viruses) would have been present and so we would expect the same percentage of cells to have become mutant in each population, no matter how many samples we removed. In fact, several (or many) samples were usually collected from one flask just to be sure to have an accurate estimate of the population size and properties in the total volume of culture. It was predicted that if the mutants were spontaneous in origin and not induced in response to the viruses, mutant frequencies would fluctuate among the different populations being sampled. Populations that happened to have no mutants in the original inoculum would show zero-growth and produce no colonies when sampled. If by chance there had been several mutant cells in an inoculum, then these cells would have flourished in the medium despite the presence of viruses, and a substantial number of descendant cells would have been removed in samples for plating and colony formation. If the mutation had occurred by chance at some time during the course of the 24-hour growth period in the virus-containing nutrient medium in the flask, then a variable number of descendant cells would have been produced, depending on the time of the mutational event in that 24-hour period. A mutant that appeared near the end of that interval would have produced fewer descendants prior to sampling, whereas a mutant that had arisen earlier in the experimental interval would have produced more progeny cells before the time when the samples were removed from the culture for plating. Thus it would be possible to choose between the two opposing theoretical possibilities by determining whether there were similar frequencies of mutants in all the populations, indicating that there had been a specific beneficial response to the external stimulus, or whether the frequencies of mutants fluctuated, which would indicate that selection acted on preexisting or spontaneous variants, if these were present at all. The results showed that mutant frequencies varied considerably from one culture to another, thus substantiating the doctrine of selection acting on spontaneous variation in populations (Fig. 7.2) and disproving Lamarckian interpretations.

There was wide acceptance of these results and of their interpretations, but an undercurrent of uncertainty continued because of the heavy reliance on statistics as the basis for the conclusions reached by Luria and Delbrück. In particular a great deal of the interpretation depended on the occurrence of samples that lacked mutants altogether. Conclusions based in large part on negative results (the absence of mutants in this case) is unsatisfactory because the failure of growth can result from any one of a hundred reasons, not just because spontaneous mutants happened to be absent from some populations. There were various attempts after 1943 to provide further validation for the neo-Darwinian principles, but direct evidence in support of the concept of selection of spontaneous mutants did not appear until 1952. One of the problems of the Fluctuation Test and other experiments designed to evaluate natural selection in microorganisms was that the mutants recovered after an experimental period remained in the presence

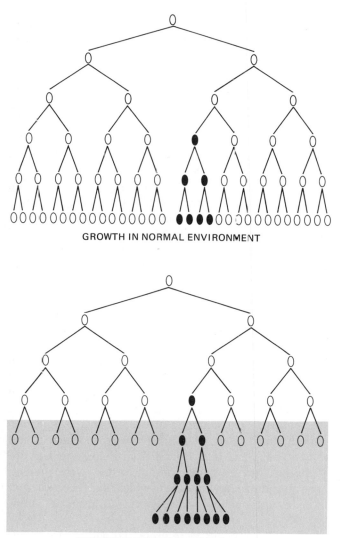

GROWTH IN NORMAL ENVIRONMENT

GROWTH IN MODIFIED ENVIRONMENT

Figure 7.2   Schematic diagram illustrating the action of selection on spontaneous mutants, leading to alterations in genotype frequencies in populations. A mutant that occurs spontaneously may be retained in a population, depending on relative disadvantage in the presence of wild type cells; but this same mutant may increase in frequency or come to predominate in a population if environmental changes alter the selection pressures.
Only the mutant cells can continue to reproduce in the modified environment. The nonmutant cells die, in this example.

of the external agent during development and reproduction. Was it even slightly possible that mutations were induced specifically but not likely to be detected by the methods and the current knowledge in our possession? The only way to resolve the matter unequivocally would be to find mutants that arose in populations lacking the specific agent—for example, to isolate virus-resistant mutants from populations containing no viruses pertinent to the mutant or the mutational event. Such data were obtained in 1952 by the Lederbergs using a replica-plating method that has since become a standard technique in modern microbial genetic studies.

In the Replica-Plating Test virus-sensitive bacterial populations were grown in a virus-free medium throughout the duration of the experiment, and samples were removed to be plated in parallel sets of growth media (Fig. 7.3). Specifically, a large amount of culture was spread on virus-free nutrient agar and permitted to grow. A replica of this virus-free growth was imprinted on a piece of velvet, which had suitable markings to orient both sets of plates for future reference, the original and the replica plates. Using the velvet on which virus-free cells had been picked up, the transfer was made to plates containing viruses in the nutrient agar. After time for these cells to grow, observations were made of the areas in which colonies did develop on the virus-containing plates. This permitted identification of the areas on the virus-free plates where the resistant cells probably would be present, but the cells on the original plate never came into direct contact with viruses. Inocula for further growth in virus-free liquid media were selected from those areas of the virus-free plates that had shown growth on the virus-containing replica plates. Continued replica plating was performed until it was possible to have few enough cells on the virus-free plates so that well-separated individual colonies could form. In this way cells from a single colony could be removed as a sampling of descendants of one presumptive virus-resistant mutant. Such cells finally were inoculated into media that were virus-containing and into a parallel virus-free set, and observed for growth. If the cells grew whether or not viruses were present, then they would represent spontaneous mutants, which had been present in the original population even though no viruses occurred in that environment.

The first time the presumptive mutants were presented with a virus challenge was at the conclusion of the experiment. At that time the observation was made that descendants of the original resistant mutants also were resistant to virus infection. This point is important because it shows that the mutants were spontaneous in origin and that the trait had been transmitted to descendant generations, far removed in time from the original cells that had been tentatively identified as virus-resistant. The mutation had been present in cells first seen in the initial population, but the experimental method led to the identification of those original mutants by virtue of the inheritance of the trait through subsequent generations.

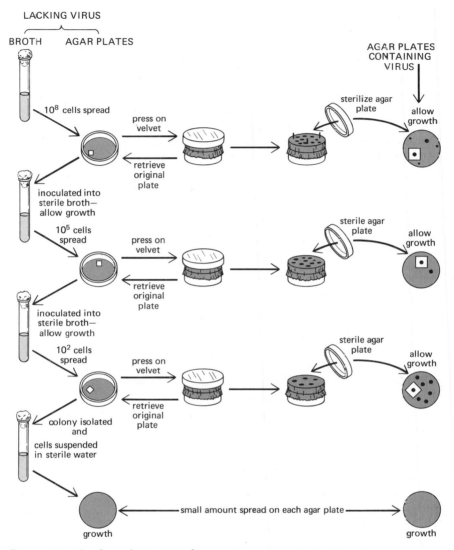

Figure 7.3   Replica plating test for spontaneous mutants. This scheme shows the design of a Replica Plating experiment, which demonstrated the spontaneous nature of mutational changes.

In the absence of the virus there was no particular selective advantage for the resistant mutants. When the environment changed, any mutants that happened to be present in the population would acquire selective advantage and become more numerous, while sensitive cells would die and leave no progeny, thus leading to a change from a sensitive to a resistant cell population. Considering the generation time of 20 to 30 minutes for such species, changes might well be observed in a single day's growth. The experimental evidence from the Replica-Plating Test, together with the substantial body of evidence collected by Darwin and subsequently by others in the twentieth century, all indicated clearly that genetic diversity arises spontaneously in populations, unrelated to the "needs" of the species, and that evolutionary change occurs as a consequence of differential reproduction of genetically diverse populations.

The same mechanisms of mutation and selection operate in all life forms and provide the foundations for continued variation and modification during evolution. These phenomena apply to viruses and bacteria, plants and animals, and the whole spectrum of life on Earth. Presumably the evolutionary process has operated in the same way since the origin of life on the planet and will continue to fashion the processions of species as long as their basic genetic properties remain as they now are constituted.

Polymorphism

In many species it has been thoroughly documented that heterozygotes occur quite commonly in populations and that the frequency of recessive alleles is not maintained by recurrent mutation alone. In polymorphic populations two or more distinct types constitute the same breeding population, such types showing discontinuous traits rather than continuous variation. If some advantage is conferred on a previously rare mutant form, or perhaps if a new and beneficial allele is introduced into a population, then the new form will increase in frequency as a function of selection. During the adjustments in frequencies of alleles and genotypes, such populations will contain substantial proportions of each phenotype and thus become polymorphic. Such polymorphism usually tends to be transient; the most probable outcome is that the newly advantageous type will predominate in the population, while the previously more common type will decrease to a level that is maintained principally by recurrent mutation. The example of the dark and light varieties of peppered moth in Great Britain and Ireland is representative of an initially transient polymorphic situation that is now balanced. The prediction is that the darker melanistic phase will predominate in industrialized areas and the light phase will be most common in rural areas. This has occurred already in some regions; moth populations in the Manchester area are essentially melanistic, whereas populations in southwestern England are of the light variety. Some individuals of the alternative type may be found in each area, but these may only represent newly mutant components of the populations. Other examples of industrial melanism have been documented quite thoroughly; each species is considered to be polymorphic only until the frequencies settle down to levels maintained by mutation and selection pressures. Many other kinds of traits, including disease resistance, have been shown to follow the pattern for transient polymorphism. In almost every case the trait that becomes more frequent in a brief interval of time, while selective advantages are changing, behaves as a dominant allele. There have been examples reported of polygenic systems or of modifying gene complexes that have been altered and thus have influenced phenotype frequencies, but the usual situation is that of a newly advantageous dominant allele that changes in frequency as a function of selection.

Balanced polymorphism occurs when selection pressures actively maintain two or more discrete phenotypes in the same breeding population, and when such populations persist for hundreds or thousands of years or longer. Some stable polymorphic systems will be established if a mating pattern encourages or enforces mating between unlike individuals. A forced mating system would be one like that described for various species of flowering plants, in which a system of incompatibility alleles is in operation. In such cases the pollen from a different individual may grow successfully in the stylar tissue of a plant and produce a fertilized egg, which develops into the next generation. Pollen from the same flower cannot function in the stylar tissue, so the incompatibility mechanism enforces cross-fertilization in such species. The different incompatibility genotypes constitute a stable polymorphic population system. Another example frequently cited is that of the pin and thrum flower types in the primrose species *Primula vulgaris*. In the pin type of flower the style is much taller than the pollen-bearing anthers; just the reverse is true in the thrum type, where the anthers are taller than the style. Pollen from the pin flowers grows very slowly on the pin style but successfully on thrum flowers, whereas thrum pollen grows only in the stylar tissue of pin flowers. Both thrum and pin types are maintained in the same breeding populations owing to the genetic pattern of allele distributions leading to the two phenotypes, and the situation is one of balanced polymorphism.

A fairly common situation in insects and other species having a short generation time is a system of balanced polymorphism based on seasonal fluctuations. One of the species of ladybird beetle (*Adalia bipunctata*) exists in a red phase and a black phase. The red phase occurs more commonly in winter and the black phase becomes more frequent during the summer, in the same breeding populations. Neither type is eliminated, because the adaptive value of each phase shifts, depending on the season of the year. A similar and well-documented seasonally-based polymorphism is representative of populations of *Drosophila pseudoobscura*, which contain distinct and different chromosomal inversion types. The relative frequencies of these inversion types change with the seasons, but the different groups are maintained by the adaptiveness of the specific inversion component relative to the seasonal temperature. The entire population is perpetuated as a consequence of the overall selection pressures; one type or another maintains the species throughout the year, but both must be present at all times in a stable see-saw frequency distribution.

In some cases of balanced polymorphism there is demonstrably superior fitness for the heterozygote relative to the two homozygous genotypes in the population. Such an example is provided in those human populations in Africa that have very high frequencies of individuals who are heterozygous for the sickle-cell allele. Even though recessives suffer the debilitations of the disease, the allele is retained in high frequency in these populations because of enhanced selective advantages of heterozygotes. Carriers of the sickle-cell trait have greater resistance to tertian malaria than do homozygotes lacking this allele. Since heterozygotes are more likely to survive the malarial infections and thus reach the age of reproduction, the allele is transmitted with high frequency in successive generations. The recessives contribute little if anything to the gene pool of these populations because death occurs during adolescence or early adulthood. The adaptive value of balanced polymorphism is clear in this particular example.

ADAPTABILITY AND ADAPTEDNESS

There are many puzzling situations in which complex adaptive systems have been more difficult to analyze in terms of specific effects of evolutionary processes than the simpler polymorphic situations mentioned above. The elaborate patterns of mimicry in animals such as butterflies, the complexes of characteristics that distinguish subpopulations of a species that inhabit strikingly different locales, the hierarchical societies of such insects as ants and bees, and the incredible morphological similarity between the female structures of some orchid flowers and the female of the bee that pollinates that kind of flower are just a few of the marvelous adaptations that have been described. Where it has been possible, genetic analysis has revealed a substantial number of genes underlying the development of such adaptive complexes. Breeding analysis and experimental studies of mimic systems have been successful enough for us to interpret a reasonable sequence of changes leading to the present arrangements of genotypes and phenotypes. Transplant experiments in conjunction with breeding analysis have shown that some 60 to 100 allelic differences may distinguish varieties of the same species of plants. Such studies also have provided evidence showing that the growth habit of the population sometimes is substantially genetic in origin and sometimes reflects responses to growth conditions in the habitats in which the populations reach maturity. We would expect that closer investigation of many adaptive complexes would also reveal the interacting factors of mutation and selection as guiding principles of the phenomena we can observe.

A distinction that we can describe briefly at this point and at greater length in another chapter is the quality of adaptability and the quantity of adaptedness. Theoretically it is possible to measure the amount of adaptedness of a population or a species, but adaptability is difficult or perhaps impossible to measure. By adaptedness we mean the status of being able to live and reproduce in a given environment, achieved by the process of adaptation. Adaptability on the other hand is the ability to adapt to a range of life situations. Adaptations, or adaptive traits, are those characteristics that lead to the status of adaptedness. The orchid species that can be pollinated only by one particular species of bee is lacking in adaptability but is highly adapted to a certain set of conditions required to perpetuate the species. If that insect species became unavailable, for whatever reason, the orchid species would lack the adaptability to survive under the new set of environmental conditions. Yet it is extremely successful in its restricted situation at the present moment, and it certainly demonstrates a marvel of adaptedness to one life style. Species that are adapted but limited in adaptability are unlikely to meet the challenges of new situations successfully, though they may be extremely successful under the restrictions to which they are closely adapted. If we consider this general dichotomy of adaptedness and adaptability in terms of short-term and long-term evolutionary advantages, we have a better basis for understanding the varieties of life styles underwritten by genetic systems. We will discuss these species characteristics in Chapter 9, after we have gained some understanding of population dynamics and models that are useful to explain an array of evolutionary case histories, some of which appear baffling at first glance.

Some Suggested Readings

Dobzhansky, Th., "The genetic basis of evolution." *Scientific American* 182 (January, 1950), 32.

Dobzhansky, Th., "Adaptedness and fitness" in *Population Biology and Evolution* (ed., R. C. Lewontin). Syracuse, Syracuse University Press, 1968, pp. 109–121.

Dobzhansky, Th., *Genetics of the Evolutionary Process.* New York, Columbia University Press, 1970.

Fisher, R. A., "Retrospect of the criticisms of the theory of natural selection," in *Evolution as a Process* (eds., J. Huxley, A. C. Hardy, and E. B. Ford). London, Allen & Unwin, 1954, pp. 84–98.

Huxley, J., *Evolution in Action.* New York, Harper & Row, 1953.

Kettlewell, H. B. D., "Darwin's missing evidence." *Scientific American* 200 (March, 1959), 48.

Mayr, E., "The nature of the Darwinian revolution." *Science* 176 (1972), 981.

Sheppard, P. M., *Natural Selection and Heredity.* New York, Harper & Row, 1960.

CHAPTER 8

GENE POOLS AND POPULATIONS

Simple mathematical models have been developed to analyze the effects of mutation pressure and selection pressure relative to each other and to the effective size of the breeding population and its gene pool, or genetic reserves. The model situation described as the Hardy-Weinberg Equilibrium is based on the brief reports published independently in 1908 by G. H. Hardy in England and W. Weinberg in Germany. At that time it was believed that alleles of a gene would be distributed in such a fashion during sexual reproduction that dominant phenotypes would occur in three-fourths of a population and recessive phenotypes in the remaining one-fourth. Misconceptions concerning Mendelian ratios were prevalent in the years immediately following the rediscovery of Mendel's work and served to weaken the acceptance of the genetic basis for evolution.

Hardy and Weinberg used the simplest model of a single gene existing in two alternative allelic forms to explain that the particular proportion of observable phenotypes would depend on the frequencies of the two alleles in a population. While this analysis is considerably oversimplified, there is no reason to expect that the model cannot be extended to systems of multiple alleles for the same gene or to polygenic systems that determine discrete phenotype classes. It is a model and as such has proven to be extremely useful in attacking the questions of genetic fluctuations in populations relative to evolutionary processes and phenomena.

The Hardy-Weinberg Equilibrium

The frequencies of two alleles of a gene and of the three genotypes these allele combinations produce can be used to evaluate components of the gene pool of a population, based partly on observations and partly on calculations. The frequency of the two alleles (we will discount the occurrence of more than two alleles for the sake of simplified discussion) would equal 100 percent (1.0). Therefore, if the frequency of one allele equals p, then the alternative allele must occur with a frequency of $1.0 - p$ (designated as q for easier reference). The combined frequencies of $p + q$ equals 1.0, so if we have some absolute value for one of these it would be a simple job of arithmetic to solve for the absolute value representing the frequency of the other allele. Such a pair of frequencies may have any absolute values, depending on the gene in question and the conditions that describe the population at any particular time. It is not at all difficult to determine the frequencies of the three possible genotypes if we know the solutions for p and q. The proportion of random fertilizations that would produce the homozygous dominant genotype would be $p \times p$, the probability that each independent factor will combine with a similar independent factor, or the product of the separate probabilities. Similarly the probability for the recessive genotype combination is the product of the separate probabilities, or $q \times q$. The conventional Punnett Square analysis shows clearly that alleles with the frequencies of p and q will yield homozygous genotypes with frequencies of p^2 and q^2 while heterozygotes will occur in the proportion of $2pq$ in this progeny population (Fig. 8.1). This relationship of $p^2 : 2pq : q^2$ is known as the Hardy-Weinberg Law or the Hardy-Weinberg Formula; it provides the basis for analyzing changes in frequencies of alleles and genotypes in the population gene pool.

In conventional breeding analyses involving a single gene difference between homozygous dominant and homozygous recessive parents, we expect to obtain the genotype ratio of $\frac{1}{4} AA : \frac{1}{2} Aa : \frac{1}{4} aa$ in the F_2 progeny generation. In those cases in which the homozygous dominant and the heterozygous genotypes produce indistinguishable phenotypes, the F_2 phenotype ratio is three-fourths dominant to one-fourth recessive, or the conventional $3 : 1$ ratio of diploid monohybrid (single gene) inheritance. However, this particular ratio of genotypes and phenotypes can be obtained only if the gametes carrying the A and a alleles occur in equal frequency and combine at random in all possible combinations at fertilization (Fig. 8.2). Any deviation from the expected ratio implies the action or occurrence of some other causative factor (or factors). In the same line of reasoning, if the values for p and q are not 0.5 and 0.5 as in the example shown in Fig. 8.2,

Figure 8.1 Hardy-Weinberg Law. Scheme showing the derivation of the Hardy-Weinberg Formula, $p^2 + 2pq + q^2$, from crosses between members of a population with alleles A and a present in frequencies of p and q, respectively.

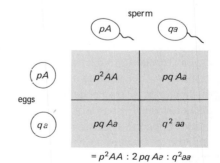

sperm

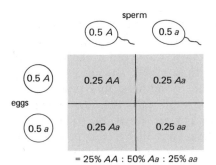

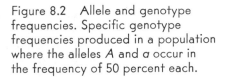

= 25% *AA* : 50% *Aa* : 25% *aa*

Figure 8.2 Allele and genotype frequencies. Specific genotype frequencies produced in a population where the alleles *A* and *a* occur in the frequency of 50 percent each.

the genotypes and phenotypes will be produced in some other proportion, depending on the specific frequencies of the gametes carrying the *A* and *a* alleles, respectively.

In the monohybrid cross in which p (the frequency of *A*) was 50 percent, or 0.5, and q (the frequency of *a*) also was 0.5, the total being 100 percent, or 1.0, we could have used the Hardy-Weinberg formula to predict the frequencies of the three genotype classes. Using $p^2 : 2pq : q^2$ we would have found $p^2 = 0.5 \times 0.5 = 0.25$; $2pq = 2(0.5 \times 0.5) = 0.50$; and $q^2 = 0.5 \times 0.5 = 0.25$. Thus we find the expected genotype ratio of $\frac{1}{4}$ *AA* : $\frac{1}{2}$ *Aa* : $\frac{1}{4}$ *aa*. If we were dealing instead with two alleles that occurred in the gene pool with frequencies of 80 percent for the dominant and 20 percent for the recessive allele, then we could figure out the expected genotype frequencies using the more cumbersome Punnett Square (the reader should try it) or the convenient Hardy-Weinberg formula, where $p^2 = (0.8)^2 = 0.64$; $2pq = 2(0.8 \times 0.2) = 0.32$; and $q^2 = (0.2)^2 = 0.04$. In this example the genotypes occur in the proportion 64 percent dominant : 32 percent heterozygous : 4 percent recessive. The specific proportions of the three genotypes found in a population obviously will depend on the specific frequencies of the two alleles in the gene pool and on the system of random matings which yield all possible combinations in the predicted ratio. Here too if the dominant and heterozygous genotypes are phenotypically identical, then the phenotype ratio would be 96 percent for the dominant trait and 4 percent for the recessive. Note that a sharp decline in the frequency of the recessive genotype in the second example (4 percent versus 25 percent in the first example) does not lead to a comparable decrease in heterozygote frequency (32 percent versus 50 percent).

One of the striking revelations of the Hardy-Weinberg equilibrium is that heterozygotes occur with much higher frequencies than one might have expected on the basis of the proportion of recessive individuals in a population. Given the constant relation between the frequencies of the two alleles and the genotypes they produce, it is possible to construct a chart or a graph in which these relations can be seen at a glance for any situation from the one extreme of 100 percent for the dominant allele to 100 percent for the recessive allele in a gene pool (Fig. 8.3).

Suppose we want to know the frequencies of a set of genotypes for a trait determined by a single pair of alleles. We could proceed by making direct counts of individuals with the recessive phenotype (or an adequate sampling of a population), since these are recognizably different from the dominant and the heterozygous phenotypes. If the heterozygotes also could be distinguished, we would have an additional parameter for interpretations. The recessive individuals comprise q^2 of the genotypes, and we can derive q, the frequency of the recessive allele, simply by extracting the square root of q^2 (the $q \times q$ component of the formula). Once we know the value for q we simply subtract from 1.0 to find the value for p, the frequency of the other allele. From p and q we can calculate the frequencies of the two other genotypes, p^2 and $2pq$, the homozygous dominant and the heterozygote, respectively.

To illustrate the method let us consider two specific examples for which values are known. The numbers have been rounded off to make the arithmetic a little easier in the explanations, but the values are approximately

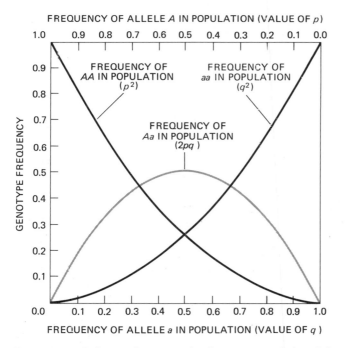

Figure 8.3 Relations between the frequencies of the alleles A and a (p and q, respectively) and the frequencies of the genotypes AA, Aa, and aa as predicted by the Hardy-Weinberg Formula.

correct for the particular populations of Americans we will discuss. Among 200 million or so Americans, about 125,000 individuals suffer from the inherited disease cystic fibrosis. The disease is expressed only in recessives, but appropriate tests can identify heterozygotes who carry one recessive allele and one normal dominant allele. If we translate the observed frequency of recessives to q^2 we find that about 125,000 out of 200 million equals 0.0625 percent, or 0.000625. The square root of 0.000625 equals 0.025, which is the value for q. Knowing that $q = 0.025$, p must equal 0.975 because the total of the two frequencies is 100 percent, or 1.0 ($p + q = 1.0$). Substituting these values for the two alleles we can calculate the three genotype frequencies as follows:

Genotype and frequency		Actual values	Actual frequencies (%)
CC	p^2	$(0.975)^2 = 0.9506$	95.06
Cc	$2pq$	$2(0.975 \times 0.025) = 0.0488$	4.88
cc	q^2	$(0.025)^2 = 0.000625$	0.06 (observed)

Since $95.06\% + 4.88\% + 0.06\% = 100\%$, we have accounted properly; but satisfy yourself of the accuracy of the arithmetic by repeating these calculations. We may note two points immediately: (1) Knowing one of the components of the formula permits us to extract the remaining information by simple arithmetic, and (2) the frequency of the *carriers* of the recessive allele, the heterozygotes, is much greater than one might have expected intuitively. About 1 in 20 Americans is a carrier of the cystic fibrosis allele, which means that the chances are high that some people in a gathering of even minimum size will be heterozygous for this trait. Since it is possible to detect carriers by examination for a slight excess of sodium in perspiration or fingernail clippings, the frequency of heterozygotes can be determined directly from population sampling as well as from Hardy-Weinberg formula calculations. The observed and the calculated values for heterozygote frequency coincide in this example. Such a correspondence can be used to interpret the population genetics of a trait and find whether heterozygotes are more fit than the homozygous dominants, as we shall see a little later.

The inherited Tay-Sachs disease, infantile or amaurotic idiocy, leads to a degeneration of the central nervous system and the death of the child between the ages of two and four. It is the result of a recessive allele that fails to produce an enzyme essential for utilization of fats; the accumulation of this material in the central nervous system leads to deterioration and death. The disease is especially common among Jewish people who trace their ancestry to northeastern Poland and southern Lithuania, although it does occur in much lower frequencies in other population groups. For the country as a whole, infants with the disease occur in 1 per 100,000 births, so the frequency of the recessive genotype (q^2) equals 1/100,000, or 0.00001, these being equivalent values. The frequency of the recessive allele in the gene pool is obtained by taking the square root of 0.00001, which equals 0.003. Substituting 0.003 for q and $1.0 - 0.003$, or 0.997, for p, we can calculate that:

Genotype and frequency		Actual values	Actual frequencies (%)
TT	p^2	$(0.997)^2 = 0.994$	99.4
Tt	$2pq$	$2(0.997 \times 0.003) = 0.006$	0.6
tt	q^2	$(0.003)^2 = 0.00001$	0.001

Here too the rarity of the recessive genotype leaves one unprepared for the relatively high incidence of heterozygotes in the population. For the country as a whole, 6 people per 1000 carry the recessive allele. In a total population of 200 million about 1.2 million individuals are heterozygous for this lethal inherited condition ($200,000,000 \times 0.006 = 1,200,000$). Heterozygotes are much more frequent in Jewish populations, as mentioned above, and about 1 in 30 (or about 33 per 1000) is a carrier in such groups in New York City. This is more than five times the frequency found in the country as a whole, which includes people of Jewish ancestry, but the heterozygote frequency for non-Jews is considerably lower than 0.6 percent.

It should be clear just from these two examples that the frequencies of monohybrid genotypes and phenotypes do not necessarily occur in the classical Mendelian ratios of 1 : 2 : 1 or 3 : 1, respectively. The relative proportions of genotypes and the phenotype expressions that they direct are a function of the allele frequencies in the gene pool. Depending on these frequencies the absolute percentages will vary, but will fit the proportion described by the Hardy-Weinberg law as $p^2 : 2pq : q^2$. The frequencies of particular alleles of a gene may vary from one population to another of the same species, as is true for the Tay-Sachs alleles just described. The difference between brown and blue eye color is a function of a pair of alleles for one gene, and brown is the dominant trait. Yet in some human populations blue eyes are extremely common, in others blue eyes never appear, and there are all gradations in between. Even though the absolute allele frequencies may vary among populations, there still could be an equilibrium situation in each of them. The condition of equilibrium does not necessarily imply an unchanging situation; rather, it indicates that the genotype frequencies are dependent on the allele frequencies and are not subject to fluctuations except as the frequencies of the alleles may change with time.

Factors that Influence Equilibrium Frequencies

The frequencies of the three related genotypes will remain constant from generation to generation only in a static evolutionary situation. Five principal factors may influence the frequencies of alleles and genotypes in the gene pool and thus affect the pattern and direction of evolutionary change in a population:

1. mutation pressure
2. selection pressure
3. number of breeding individuals in a population
4. system of mating or breeding
5. migration of genes (gene flow)

We will consider these factors in turn.

MUTATION AND SELECTION PRESSURES

The mutation pressure may be equated with the rate of mutation, and the probability of a mutation's occurring varies from gene to gene and from one species to another. Because of this, we will consider the average mutation rate to be about 0.0001, that is, about 1 per 10,000 gametes will carry a mutant allele for a particular gene locus. Since the continuity between generations is a function of sexual or asexual reproduction, the allele frequency in the gene pool is relevant only to the cells that initiate the next generation. Mere presence in the gene pool is insufficient for effective evolutionary dynamics of the allelic potential that resides there. The following illustration will clarify the distinction between actual and effective size of the gene pool in relation to a particular pair of alleles.

Suppose that we observe the birthrate of a particular recessive genotype to be 1 per 10,000 individuals, or 0.0001 as the frequency of new recessive genotypes added to the gene pool. In this example we will consider the trait to be lethal, that is, that these recessive individuals do not reproduce because they die before reaching the age of reproduction or are sterile and incapable of reproduction regardless of their longevity. This constitutes a genetic death since the genotype cannot perpetuate itself. Although the recessives cannot contribute to the gene pool of the next generation, new recessives will continue to be produced by heterozygotes in the population. Interbreeding between heterozygotes, as in the situation of $Aa \times Aa$, will generate all three genotypes, including the recessives (see Fig. 8.1). Using the Hardy-Weinberg formula we find that $q^2 = 0.0001$, and $q = \sqrt{0.0001} = 0.01$. The recessive allele a thus is present in 1 out of 100 gametes in the population. From $1.0 - q$ we find that $1.0 - 0.01 = 0.99$, and that $p^2 = (0.99)^2 = 0.9801$, or 98.01 percent of the population is homozygous for the dominant normal allele. The heterozygote frequency is obtained from $2pq = 2(0.99 \times 0.01) = 0.198$, or 1.98 percent, whereas q^2 was observed to be 0.01 percent, and the same value is obtained from the calculation $q^2 = (0.01)^2 = 0.0001$, or 0.01 percent are of the recessive genotype. Although the frequency of recessives is 1 per 10,000 individuals, almost 2 percent are heterozygotes; 1 in 50 members of the population is a carrier of the lethal trait.

For a lethal trait the mutation rate can be obtained directly from the birthrate of recessives, since the general equation states that the frequency of recessives is a function of the mutation rate (u) and the selective disadvantage (s), or $q^2 = u/s$. Where every recessive genotype dies without reproducing, $s = 1.0$; that is, the selective disadvantage for the genotype is 100 percent. If $s = 1.0$, then the frequency of recessives equals the mutation rate, according to the equation shown above. Mutation rates for dominant to *lethal* recessive alleles ($A \rightarrow a$) generally are obtained directly from the frequency of recessive genotypes in the population, or from their birthrate, since these values are equivalent.

Now suppose that the selective disadvantage is reduced, perhaps by some change in the physical or biological environment, so that only 10 percent of the recessives fail to reproduce and 90 percent, on the average, can contribute to the gene pool and so the next generations. If only 10 percent of the recessives born do not contribute, then q^2 increases in the population, as we can see by substituting the new values in the general formula:

$$q^2 = \frac{u}{s} = \frac{0.0001}{0.1} = 0.001, \text{ or } 0.1 \text{ percent}$$

The mutation rate is unaffected by the altered selective disadvantage of course, but there now is 1 recessive per 1000 individuals in the population, all potential contributors to the gene pool. From the altered value for q^2 we now derive the new frequency for the recessive allele, q, to calculate the proportions of homozygous dominant and heterozygous individuals produced from the gene pool. We find that $q^2 = 0.001$ from our observations, so that $q = \sqrt{0.001} = 0.032$; $p = 0.968$, so that $p^2 = (0.968)^2 = 0.937$, or 93.7 percent; and for the frequency of heterozygotes, $2pq = 2 (0.968 \times 0.032) = 0.062$, or 6.2 percent of the population. To summarize, in this population there would be 93.7 percent homozygous dominant, plus 6.2 percent heterozygous, plus 0.1 percent recessive members. With the altered selective disadvantage for the recessive genotypes their frequency increases, of course, but heterozygotes now represent about 6 percent of the population, or 1 in 16 individuals carries the recessive trait whereas only 1 in 50 was a carrier in the earlier situation.

It should be clear from this example that changes in the frequencies of alleles are more likely to occur as a function of selection pressure than of mutation pressure, since selective disadvantage modulates the values for p and q substantially. From the arithmetic we also should deduce that selection would proceed very slowly if p, q, or s were very small, and would not occur at all if any of these quantities equaled zero. Substitute appropriate numbers in the formula $q^2 = u/s$ and satisfy yourself that these conclusions are valid. The generalization that emerges from these considerations is that evolution relative to a particular gene in a population depends on one allele's having a selective advantage over its alternative allele. At one time some sets of alleles were believed to be selectively neutral, no one of them having an advantage relative to another alternative at the locus. But recent data have brought this idea into dispute; for example, the A-B-O major blood group factors in humans were considered to be equivalent in selective advantage, but current evidence showing a higher incidence of traits such as stomach cancer and duodenal ulcer in one or another of these genetic backgrounds has prompted a more careful review of the concept of neutral alleles. The interpretations are somewhat controversial at present.

On casual inspection it might seem that the elimination of an allele with a selective disadvantage of 100 percent (lethal) would be a fairly rapid process because none of these lethal genotypes can reproduce. This actually is not the case, as the following hypothetical situation will show. For purposes of illustrating the point, suppose we start with a population in which all the individuals are heterozygous. The frequency of each allele would be 50 percent because each individual is genotypically Aa and produces gametes carrying A and others carrying a with equal frequency. We thus start with allele a as 50 percent of the gene pool. On interbreeding the heterozygotes to produce the next generation we find that the three genotypes occur in the expected frequency of 25 percent AA : 50 percent Aa : 25 percent aa, as can be determined using the Hardy-Weinberg formula.

The birth frequency is not the same as the frequency of the genotypes that will produce the gametes for the next generation, however; the aa genotypes die and only the AA and Aa genotypes reproduce. To calculate the genotype frequencies for generation number 2, we must adjust the values, since one-third of the breeding population is now AA and two-thirds are genotypically Aa (but they were 25 percent and 50 percent of the births). With adjusted genotype frequencies of $\frac{1}{3} AA$: $\frac{2}{3} Aa$, the frequencies of the two alleles now are calculated to be two-thirds A and one-third a in the gene pool, since all the gametes produced by AA are A, and of the Aa gametes half are A and half are a. The frequency of the a allele has decreased from 50 percent to 33.3 percent in just one generation! The genotype frequencies of the third generation can be calculated from p^2 : $2pq$: q^2 as follows:

$$(\tfrac{2}{3})^2 AA : 2(\tfrac{2}{3} \times \tfrac{1}{3}) Aa : (\tfrac{1}{3})^2 aa = \tfrac{4}{9} AA : \tfrac{4}{9} Aa : \tfrac{1}{9} aa$$

Again the aa genotypes are eliminated, and the frequencies of alleles in the *effective* gene pool become $\frac{3}{4} A$: $\frac{1}{4} a$ since the frequencies of the two genotypes in the population become $\frac{1}{2} AA$: $\frac{1}{2} Aa$. Another striking reduction has occurred; the a allele now is only 25 percent of the effective gene pool, half as many as were present in the first generation. If we continue to use the Hardy-Weinberg formula to calculate the allele frequencies in each subsequent generation, always altering the frequencies of A and a to conform with the elimination of the aa genotypes, we find that the elimination of a becomes less and less efficient in each succeeding generation. In fact, after ten generations allele *a* represents 9 percent of the effective gene pool, and by the hundreth generation the *a* allele comprises 1 percent of the gene pool.

The main reason for the reduced effectiveness of the same stringent selection pressure is that as fewer *aa* genotypes are produced in each generation, a larger proportion of the *a* alleles of the gene pool are generated by the heterozygotes. Not only are the *a* alleles protected from selection in the heterozygote (selection acts on the phenotype expressed by the *aa* genotype and not on the gene itself), but there are fewer and fewer heterozygotes in each generation, too. In diploid species, variability persists in the gene pool as a function of heterozygous carriers which replenish the alleles lost as recessive genotypes, even when there is 100 percent elimination of the recessive genotypes in each generation. But this replenishment becomes reduced gradually and regularly. Under most conditions the allele ultimately comes to be maintained in the gene pool by recurrent mutation in the case of a lethal recessive trait. For this reason we may compute the mutation rate for a recessive lethal allele directly from the birthrate, as in the formula $q^2 = u/s$. For any detrimental allele with a selective disadvantage of less than 100 percent, the reduction in its frequency is slower still. The lower the disadvantage, the slower the action of selection on the rate of elimination of such genotypes once they are expressed in recessive individuals (Fig. 8.4). In such cases the surviving recessives contribute to the effective gene pool along with the more numerous heterozygotes carrying the allele.

SELECTIVE ADVANTAGE IS RELATIVE

If we examine some particular set of genotypes in one kind of environmental situation we may find a selection pressure operating that is rather different from that in some other external background. Since selection pressure leads to differential reproduction of diverse genotypes, we can explore this phenomenon of evolutionary potential by experimental analysis. A convenient test system was provided by *Drosophila* stocks, which were crossed in such a way as to produce genotype classes whose predicted and observed proportions could be compared in different environmental backgrounds. In a conventional backcross experiment, heterozygous flies were interbred with homozygotes carrying recessive traits. The predicted ratio for the two progeny classes in a monohybrid backcross experiment would be 1 : 1 if both genotypes were entirely equivalent. One advantage of the backcross is that each genotype progeny class has a different phenotype, so we can recognize what alleles are present simply by inspecting the phenotypes that are produced. In this set of breeding tests we follow the relative success of the recessive allele responsible for miniature wings (*m*) in a genotype containing either a pair of wild type alleles for long bristles (*BB*) or recessive alleles for bobbed bristles (*bb*). As shown in Table 8.1, the

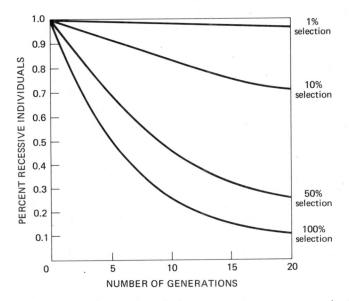

Figure 8.4   Individuals with the recessive genotype are eliminated more and more effectively as the intensity of selection increases. Beginning with a population containing 1.0 percent recessives, complete selection reduced their frequency to about 0.1 percent; less dramatic changes prevail at lower selection pressures over this same span of time.

three crosses should yield equal frequencies of the two progeny classes, which resemble the immediate parents in each case, unless there is a selective disadvantage for one of the two classes. Since we expect equal proportions of progeny types, the numbers of offspring can conveniently be tabulated as percentages of recessives against 100 percent as the standard for the wild type heterozygote class. In other words if we recover 100 heterozyotes and 100 recessives then we have 100 percent of the expected recessives in the backcross progeny. If instead we obtain 100 heterozygotes and only 80 recessives, then we tabulate that 80 percent of the expected numbers of recessives actually were produced in the cross. The breeding populations were raised in six different environments, which varied in temperature and relative crowding; the control populations were raised under optimum conditions for *Drosophila,* crowded cultures maintained at a temperature of 24° to 25°C. The results of the experiments illustrate some useful points:

1. The selective disadvantage varies for a particular mutant phenotype as a function of the surroundings in which the population lives. The miniature-winged flies are more successful when the temperature is lower than

TABLE 8.1 EFFECTS OF ENVIRONMENTAL FACTORS ON GENOTYPE FREQUENCIES*

Genotype	Normal conditions	Temperature			Population Density		
		15–16°C	24–25°C	28–30°C	Sparse	Crowded	Very Crowded
mmBB	70 ± 0.9	91	69	64	93	69	47
MMbb	85 ± 0.8	75	85	94	77	85	92
mmbb	97 ± 0.3	—	—	—	—	—	—

* Expressed as percentage actually observed relative to the expected 100 percent per class.

the optimum for wild types or when conditions are less crowded than preferred by wild types. The bobbed-bristle mutants fare better at temperatures above the optimum or under very crowded conditions. Neither of these mutants proves to be as successful as the wild type, however, in either the normal conditions for wild type or in one of the less optimum environments. The miniature-winged progeny therefore have a greater disadvantage at higher temperatures or very crowded conditions, and the bristle mutants are at a disadvantage in populations that contain wild type flies under most of the conditions, but especially at lower temperatures or in sparse populations.

2. The same genes may have a different selective disadvantage under identical environmental conditions, depending on the remainder of the genotype of the individuals. Thus the recessive *mm* is produced at only 70 percent of the expected frequency in a genotype also containing *BB,* but *mm* occurs at approximately the same frequency as the wild type heterozygotes in a genetic background that includes *bb.* The analysis of selection pressure is complicated by the many levels of interaction of the particular alleles under study, but the important point is that selection is relative to all the components of the life situation of a population. These components include the elements of the gene pool as well as the physical and biological factors in the external environment.

A well-documented example of relative selective disadvantage concerns the alleles for normal and sickle-cell hemoglobin. Individuals who are recessive for the sickling allele have a shortened life-span; thus the allele has a selective disadvantage of less than 1.0, since some of the recessives contribute to the gene pool. Heterozygotes often have a mild anemia, which probably is of little consequence in their daily lives. The alleles for normal adult hemoglobin (Hb^A) and sickle-cell hemoglobin (Hb^S) can be detected by simple blood tests for each of the three genotypes in the several parts of the world where the sickling allele occurs. There are some regions in Africa in which the incidence of the heterozygotes may be as high as 40 percent and as many as 4 percent of such populations may be recessives, although the frequencies usually are lower than this. Ancestors of Americans of African descent were brought from various parts of that continent during the shameful years of the slave trade, and it has been estimated that about 22 percent of such individuals were heterozygous for the Hb^S allele. Present-day Afro-Americans include some 9 to 10 percent heterozygotes and about 1 in 400 individuals suffers from sickle-cell anemia, having the recessive genotype $Hb^S \, Hb^S$.

Two questions among others should be asked immediately: (1) Why is there such a high incidence of a demonstrably detrimental allele in some African populations? And (2) what has contributed to the decline in frequency of the Hb^S allele during these past several centuries since blacks were brought to the American continent? The answer to the first question has been documented quite convincingly. Comparisons of different African regions and their populations have revealed a correlation between the prevalence of tertian malarial infection and a high frequency of the sickle-cell trait. In this case, as with several other genetically determined hemoglobin defects, the malarial parasite does not set up a virulent infection in the red blood cells of individuals having some abnormal hemoglobin, whereas people whose hemoglobin is entirely of the normal kind are subject to these usually lethal parasitic infections. In regions where there is a high incidence of malaria and of death resulting from the disease, heterozygotes are more likely to survive the infection and live to the age of reproduction than are homozygous normals. Despite the severe disadvantage of the allele in the recessive members of these populations, there is a high adaptive value for the allele in the heterozygotes and through them for the whole population. The high frequency of the Hb^S allele in such gene pools is a function of the selective advantage in the heterozygotes who, because they survive the depredations of malaria, make substantial contributions to the gene pool. It is possible to show such an advantage for the heterozygotes by using the Hardy-Weinberg formula to calculate the frequencies of the three genotypes from blood samples taken directly from the populations. If the three genotypes are in equilibrium and gene frequencies are unchanging, then the observed frequencies should equal the predicted frequencies calculated from $p^2 : 2pq : q^2$. Instead of this the observed proportions of the three genotypes show more heterozygotes than expected, thus indicating the occurrence of a selective advantage for heterozygotes relative to the homozygous normals. Such populations exist in a balanced polymorphic situation inasmuch as the $Hb^A \, Hb^A$ and the $Hb^A \, Hb^S$ genotypes are maintained in a relatively stable proportion as a function of selection pressures operative in different populations and environments.

The answer to the second question should be obvious from the preceding discussion. Heterozygotes do not enjoy a particular selective advantage in malaria-free areas; thus the detrimental nature of the sickling allele is made more manifest and heterozygotes, having no advantage relative to homozygous normals, decline in frequency. Based on the estimated frequency of heterozygotes in initial populations and the mortality rates for recessive individuals over a span of about 12 generations, the predicted heterozygote frequency for American blacks is between 9 and 10 percent. This in effect is the observed frequency (Fig. 8.5). The occurrence of an equilibrium frequency distribution means that there is no advantage of $Hb^A HB^S$ over $Hb^A Hb^A$ genotypes; it does not mean that further changes in allele frequency will not occur. Indeed, the substantial modifications noted over a limited number of generations provides a remarkable demonstration of evolution in action. The gene pool has changed as a function of selection pressure, which is the crucial feature of evolutionary change with time. The frequency of the sickling allele still is quite high; more than 2 million American blacks carry the sickle cell trait and about 1 in 400 or perhaps 50,000 to 60,000 people suffer from this blood disorder.

SIZE OF THE BREEDING POPULATION

From various mathematical calculations based on the estimated rate of respective fixation and loss of the two alleles of a gene, relative to average rates of mutation and selection, a generalization has been derived concerning effective population size for evolutionary phenomena. Three sizes of breeding populations are (1) large, containing more than 25,000 breeders, (2) small, having fewer than 250 breeding individuals, and (3) intermediate, with from 250 to 25,000 individuals contributing to the gene pool.

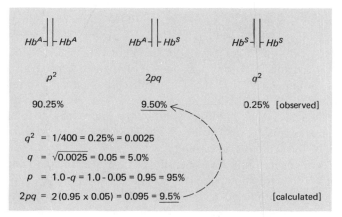

NO SELECTIVE ADVANTAGE FOR HETEROZYGOTES

Figure 8.5 Hardy-Weinberg equilibrium example. Correspondence between the calculated (predicted) and observed percentages of individuals who are heterozygous for the sickle-cell allele indicates that there is no selective advantage for heterozygotes among present-day Americans of African descent.

Because of the effects of chance alone, there is a fairly high probability that one allele may be lost; thus in small populations the alternative allele can become fixed, despite mutation and selection pressures. The rate of loss of heterozygosity can be calculated to be greater than either mutation rate or selective disadvantage in such situations, which are described as undergoing *genetic drift*. Because of drift an advantageous allele may be lost or a detrimental allele may become established in the gene pool purely by chance. Since allele frequencies may fluctuate according to chance rather than selection, evolution in small populations may not necessarily be in the direction of improvement. The principle of genetic drift has been invoked to explain various examples of nonadaptive traits in populations, these presumably having become established by chance at a time when the population was small. If there happens to be no mutation to restore the lost advantageous allele of the gene, a population may possess some nonadaptive trait even after reaching a larger size, one more favorable for evolutionary progress. There is little agreement among evolutionists on the significance of genetic drift in shaping population characteristics, although the theory was presented over 40 years ago by Sewall Wright. Many believe that the forces of selection are so great that little influence could be exerted by chance alone in any population. It is admitted, however, that some traits may become widespread in a population when the population is initiated, as a function of the first members of the group's being genetically alike for some specific feature. The *founder principle* states that some traits lacking distinctive adaptive value (but *not* being nonadaptive) may become fixed or very widespread either because the few founders of that population happened to have some trait in common or because only a few individuals were the ancestors of most of the members of the later group. The descendants of the founders, even in larger populations, might retain such traits despite lack of any particularly strong advantage for their retention at a later time. In effect the influence of chance is admitted, but only for traits that seem insignificant for the well-being and success of the eventual population. The power of selection in guiding evolutionary improvement is of such magnitude that attempts to reduce its effectiveness are received with skepticism by much of the scientific community.

The problems of endangered species generally can be attributed to the increased probability for extinction when there are few breeding individuals or too few progeny produced to perpetuate the species. According to some views there is an additional problem if traits become fixed in such small populations by genetic drift and if such traits happen to be disadvantageous in any degree. The wholesale depletion of some species of whales leads to predictions of extinction because of the reduced chances of their finding mates in the ocean vastnesses. It is not considered to be endangered because of genetic drift phenomena, in the view of those biologists who are most concerned about the deliberate destruction by the human species of global natural resources.

Unlike the small population, large populations are considered unlikely to evolve toward improvement because mutation and selection pressures operate at levels too low to be effective. The large population is believed to be in a static situation because exceedingly long times would be required for mutation and selection pressures to influence changes in allele frequencies significantly. Such populations undergo little or no effective evolutionary change because of the slow responsiveness of the gene pool to environmental fluctuations.

The intermediate-sized population is the type most likely to evolve in the direction of adaptation, because the rates of mutation, selection, and chance alteration of allele frequencies are balanced forces operating on the gene pool. Such populations appear to be most responsive to changing conditions of the genotypes, phenotypes, and the environment.

Clearly, however, many species consisting of large breeding populations (more than 25,000 breeding individuals) are quite successful in an evolutionary sense. One answer to this apparent contradiction is that such groups exist as compartmented enclaves. Although in theory interbreeding may occur between any members of a single species, in fact effective access to partners is provided within a compartment more often than between separate compartments. Pockets of breeding groups thus become established and gene exchange is not completely randomized throughout the species, no matter what its total population may be. One would expect different mutations to occur in different compartments simply because these are chance events, and one would expect different selection pressures to prevail in different regions of the total distribution area. Given such variation among compartments of a population, it would be reasonable to expect that maximal evolutionary opportunity would be realized, since each enclave is akin to an "experiment" in evolution. Because there is some gene flow between compartments as a function of occasional interbreeding between members from different groups, favorable alleles can be spread fairly rapidly as selection pressures change. Genotypes that might have some disadvantage in one compartment may be stored in other compartments as recessive alleles in heterozygotes, or even may be advantageous under a different set of selection pressures elsewhere in the species range. Thus while the total species population may considerably exceed 25,000 breeders, the effective breeding groups may be intermediate in size and therefore favorably disposed for efficient and rapid evolutionary change.

MIGRATION OF GENES (GENE FLOW)

As we have just mentioned, new diversity may be introduced into one gene pool by occasional interbreeding between members of different compartments of a species population. Enhanced variability leads to increased probabilities for favorable genetic modification in evolution. A great deal of evidence has come from studies of hybridization involving members of different species or races of a species, as well as from considerations of "hybrid vigor" in stimulating the genetic structure of a population.

There are many examples of genetic modification of populations as a result of gene flow between groups subsequent to the opening of new environments and changes in selection pressures thus engendered. If a species has diversified genetically so that some populations live more successfully in a moderately wet habitat and others exploit moderately dry areas, then interbreeding is rare simply because such group encounters are infrequent at best. If the habitats change so that many intermediate situations come into existence, then genetically intermediate forms can become established from the progenies of such occasional interbreedings. Few such progeny of two differentiated populations would be expected to become successfully installed if the environment contained principally either wet or moderately dry habitats, as in our example. After all, the successful parent populations had become adapted to each of the distinct habitats and any genetically intermediate forms probably would be less successful than either of the established parents in well-defined environments. But if the habitat were to be disturbed in some way, then varying portions of the species range might be modified such that intermediate areas would come into existence. Human activities often are responsible for disturbance of habitats, whether through agricultural practices, engineering, denuding forests and woodlands, or other ways of making the land conform to our purposes. Once there is a greater variety of living conditions instead of exclusively wet or dry living zones, populations may encounter each other more frequently, producing more hybrid progeny. Such hybrid offspring may be better suited to intermediate sorts of living space and may have more chance to become established than either parent population, the latter being more suited genetically to restricted living conditions. Edgar Anderson first pointed out the effectiveness of such "hybridization of the habitat" about 25 years ago. His observations were substantial aids to a better understanding of changes in population structures resulting from new habitats and new opportunities for genetically diverse groups to become established in areas that were unsuited to the parent types. Gene flow is not restricted only to cases of habitat disturbances; in most populations there is sufficient gene exchange from occasional interbreedings to permit management of the maximum effective potential of the total gene pool of a species.

SYSTEM OF MATING OR BREEDING

The genotype diversity of populations is modulated by the degree of inbreeding or outbreeding that occurs or is permitted. Inbreeding leads to homozygosity and consequent rapid exposure of alleles to the influence of selection on the expressed genotypes. There must be compensating mechanisms that operate in obligately self-fertilizing species, since many plants and some animal species apparently are successful despite highly homozygous genotype frequencies. Inbreeding tends to have more drastic effects in species that regularly outbreed. A familiar example of such consequences is provided from studies of breeding practices used to obtain vigorous strains of hybrid corn. Corn (*Zea mays*) is a highly heterozygous species, which undergoes regular cross-fertilization to produce progeny. In developing parent strains to be used eventually to produce hybrid corn seed, plants are artificially inbred to yield highly homozygous lines. Such homozygous strains generally are not very hardy and many cannot be used because of their poor growth qualities. Some can be used for the production of hybrid seed, and different homozygous parent strains are interbred to produce extremely vigorous, high-yielding, heterozgous offspring. In this species the effects of inbreeding and increased homozygosity are startling in comparison with cross-fertilizing strains that usually produce vigorous plants.

Outbreeding usually leads to increased heterozygosity and therefore to increased levels of genetic diversity in populations. Since selection acts on diverse genotypes, we would expect such mating systems to have substantially greater potential for evolutionary success than would be true for inbred species. There are numerous examples of adaptations that ensure outbreeding in species. The separation of the sexes as different individuals, whether in plants or animals, is a virtual guarantee for increased heterozygosity, because cross-fertilization must occur to produce the next generation.

Among plants some truly remarkable mechanisms function to ensure cross-pollination whether or not the eggs and sperm are produced in the same flower or individual. A common device is one in which the pollen-bearing anthers mature at a different time from the receptive structure in the female organ of the same flower. Such flowers usually are pollinated by insects; when the insect visits a flower it may rub against mature anthers, thus picking up pollen. If the insect lands on another flower with a receptive style and immature anthers, the pollen may be transferred to that style and thus effect cross-pollination; the subsequent cross-fertilization takes place when the sperm from the pollen reaches the egg cell after its growth through the stylar tissue. This process is repeated as the insect continues to ply its way from flower to flower in a day's work. In the interesting example of the Dutchman's Pipe species of *Abutilon,* the insect becomes trapped in the flower after it has entered the chamber formed from the petals. The insect delivers pollen (from a different flower it had visited earlier) to the receptive female structure, and fertilization occurs a few days later. At this time the flower more or less collapses, but the anthers have become mature in the meantime. The insect now can escape from the confinement of the flower, but it picks up some of the pollen from the mature anthers and flies off to another flower to repeat the cycle.

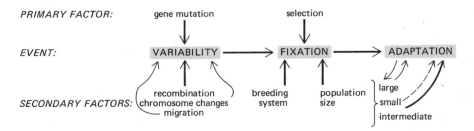

Figure 8.6 Summary of interrelationships among evolutionary events relative to phenomena that affect these events.

Summary of Interrelationships

Variability (diversity) is produced by more than one process. New genetic material arises by mutation; new genotypes may be generated also from recombination, gene flow between differentiated populations, or changes in chromosome number or structure. The incorporation of such diversity into the gene pool depends on selection as the principal guiding force by which variation becomes fixed in the genetic structure of a species or its populations (Fig. 8.6). Effective selection pressure depends on both the population size and the operative breeding system. Adaptation, as evolutionary improvement, is more or less probable according to the size of the breeding population. Equilibria are likely to be established in very large populations so that a relatively constant level of variability is maintained, and selection is less effective. In small populations genetic drift may operate to fix alleles in the gene pool according to chance; hence such groups may or may not experience adaptation. Also, because there are few breeders, favorable genotypes may not be produced among the few combinations that are present in their progenies. Evolutionary improvement is most likely to occur in the intermediate-sized population or in compartments of intermediate size in larger groups. Evolutionary improvement is most probable in these cases because selection is most effective and the increases in frequency of favorable alleles and genotypes provide the framework for adaptation.

Some Suggested Readings

Dobzhansky, Th., *Genetics of the Evolutionary Process.* New York, Columbia University Press, 1970, pp. 95–125, 230–266.

Grant, V., *The Origin of Adaptations.* New York, Columbia University Press, 1963, pp. 141–300.

Waddington, C. H., *The Strategy of the Genes.* London, Allen and Unwin, 1957, pp. 109–140.

Williams, G. C., *Adaptation and Natural Selection.* Princeton, Princeton University Press, 1966, pp. 20–55.

CHAPTER 9

EVOLUTION
OF
GENETIC SYSTEMS

The genetic system represents the totality of genetic mechanisms and controls operative in the species. Included within the system would be processes such as mutation, recombination, development of genetic potential, mechanisms of regulation, and other components directly pertinent to the genetic material. Since all these component parts of the genetic system are subject to heritable variation, we can expect the entire apparatus to be subject to the force of natural selection in fixing adaptive diversity in the gene pool. If mutation and selection pressures indeed shape the genetic system, as seems reasonable to infer from available evidence, then the genetic material that modulates and underwrites the system also would undergo evolutionary changes with time. We should thus expect that modifications in the genetic system would be as reflective of evolutionary change as is the case for the usual gene frequencies that are studied. We should be able to predict that some adaptations would be related directly to the genetic system itself as an outcome of mutation and selection, precisely as we find adaptations for such standard phenotypic traits as anatomy and physiology.

The major function of the genetic system appears to be related to factors that provide a compromise between *immediate fitness* (requiring genetic constancy) and *flexibility* (requiring genetic variability). This general concept has emerged from many genetic studies, but especially from intensive analysis by combined genetic and cytological methods. The heyday of these studies in the 1930s and 1940s provided us with a *cytogenetics* perspective, which has enjoyed a recent revival in molecular genetics circles. Because the features of genetic constancy and flexibility contrast with each other, we would expect to find compromise situations that permit both features to be accommodated within a single system having even greater adaptive value.

The constancy required for immediate fitness may seem inadaptive on first inspection, since populations with this kind of genetic system would contain relatively few genotypes. In any situation subject to changing selection pressures there would be disadvantages for a population of limited genetic diversity because the few genotypes might all prove unsuitable under new sets of conditions. Genetic variability in a flexible genetic system leads to the production of many kinds of genotypes, some of which may be adapted to new conditions while others may be less suited or unsuited. The significance of flexibility is that many possibilities are produced from a highly diverse gene pool, and the chances are pretty good that some of the genotypes will prove successful. In sexually reproducing species new genotypes must be generated and existing genotypes must be perpetuated if the species is to continue. The requirements for constancy and for variability generally are intertwined. Different kinds of genetic systems manage to balance the short-term advantages of constancy with the long-term advantages of flexibility. The means for achieving such balance and the degree of the balance, however, vary among genetic systems.

Recombination as a Component of the System

Within the genetic system there exist subsystems related to particular genetic functions, including mutation and recombination among others. The recombination subsystem has been studied extensively, and we will discuss only this component to illustrate relevant principles. In sexually reproducing eukaryotes the recombination events depend on both meiosis and fertilization. Genes and groups of genes segregate during meiosis; reassortment or recombinations of linked genes occur at fertilization. The amount of recombinational potential actually realized is a function of the degree of segregation permitted within the genetic system, that is, of the meiotic events and the nature of the chromosomes that undergo meiosis. Because genes exist in linkage groups, all genes do not reassort at random relative to each other; hence only a fraction of the recombination potential will be realized. Linked genes tend to stay together more often than not, whereas genes on different chromosomes (in different linkage groups) follow the Mendelian Law of Independent Assortment (see Fig. 5.5).

Many species have been studied and a range of recombination potentials has been found, from the one extreme of completely lacking, as in most asexual forms, to the opposite end of the spectrum in which little or no restriction is imposed on reassortment other than that imposed by the genes' being linked together in chromosomes. There are two principal considerations of a recombination subsystem relative to the variation that is generated in this fashion: (1) the *amount* generated, and (2) the *evenness of the flow* of such generated variability. In some restricted systems the flow is sporadic; bursts of variability may be separated by longer intervals of genetic constancy in populations. The more restricted the system, the longer the period during which there is genetic constancy. The least restricted recombination subsystems are those that release a slow but steady trickle of variability all or most of the time.

Various factors influence recombination at the stage of meiosis, but the two that are most significant are the *chromosome number* for the species and the frequency of *crossing-over* between linked genes. We will consider each of these factors and then move on to factors that intervene at the stage of fertilization.

If each chromosome pair in the diploid nucleus contained just one pair of alleles in the heterozygous state, that is, A/a on one chromosome pair, B/b on another, C/c on yet another, and so forth, then the number of genetically different *kinds* of gametes produced at meiosis would equal 2^n where n equals the number of chromosomes in a haploid set. As one more pair of chromosomes is added to the nucleus, the number of genetically different kinds of gametes is doubled, if this pair also is heterozygous for just one pair of alleles. Thus, for 5 chromosome pairs there would be 2^5 kinds of gametes, or 32; for 6 pairs of chromosomes bearing one pair of heterozygous alleles on each chromosome pair, the gamete variety would be 2^6, or 64; 2^7, or 128, kinds of gametes would be produced for 7 chromosome pairs; and so on. It obviously follows that reduction by just one chromosome pair will reduce gamete genetic variety by one-half, and complete homozygosity for any one pair of chromosomes in the set would have the same effect. Many flowering plant species are polyploid, having more than the two chromosome sets of the usual diploid. When the chromosome number is doubled from the diploid to the tetraploid condition, there is a substantial increase in potential diversity because more pairs of heterozygous alleles may be present on the chromosomes. If there were 7 pairs of chromosomes in the diploid nucleus and this doubled to the tetraploid state with 14 pairs, then the number of genetically different kinds of gametes (heterozygous on each chromosome pair for only *one* pair of alleles) would increase from 128 (2^7) in the diploid to 16,384 (2^{14}) in the tetraploid population.

Reduction in chromosome number has occurred rather frequently during the evolution of some species groups among both plants and animals. Such reductions usually do not involve losses of chromosomes. Instead the translocation process more commonly known as centric fusion (see Fig. 6.10) leads to the retention of all or most of the genes which now are situated in the same chromosome rather than in seperate chromosomes. Excellent evidence for centric fusion leading to reduced chromosome number has been found in various species complexes among insects (see Fig. 6.9), and also in a variety of more highly evolved animals and plants. Since fewer genes assort independently after the chromosome rearrangements have become established, fewer genotypes would be produced than under the previous condition. Recombination of genes in the same linkage group requires a prior crossing-over event, whereas independent assortment of genes on different chromosomes is an expected outcome of meiotic segregation patterns.

FREQUENCY OF CROSSING-OVER DURING MEIOSIS

Studies of meiotic chromosome preparations have revealed consistent patterns of recombination frequency for many species. Such information can be obtained from breeding analysis in the laboratory, and also from the particular configurations assumed by chromosome pairs during the early stages of the first division of meiosis. Chromosome analysis can be interpreted in terms of recombination potential, using either laboratory or natural populations of sexual species. Chiasma figures (Fig. 9.1) generally are believed to indicate crossing-over events; thus a recombination index can be derived from calculations based on chromosome number and number of chiasmata per chromosome pair for any species. Those species that show fewer chiasmata are considered to have a greater restriction on recombination by crossing-over and thus to generate reduced variability and genotype combinations. The recombination potential can be expressed adequately by the recombination index together with the chromosome number.

Recombinant frequencies for linked genes may be reduced through more than one mechanism, according to the experimental evidence. Fewer recombinants may be produced in populations containing specific alleles that regulate crossing-over frequency, usually by limiting the chromosome exchange mechanism in some way. Recombination frequency may be reduced by the presence of chromosome inversions in polymorphic populations. In such populations different chromosome regions have experienced inversions, with a consequent high incidence of inversion heterozygotes. The gametes from such inversion heterozygotes tend to be fewer in kinds, as a consequence of duplications and deletions that arise during meiosis, which

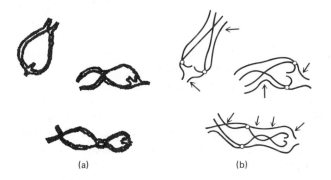

(a)   (b)

Figure 9.1   Drawing of chromosomes from a nucleus in a late portion of the prophase stage of the first division of meiosis. Chiasma locations are indicated at the arrows. There are four strands per structure since each of the two homologous chromosomes has duplicated prior to pairing and crossing-over, but note that only two of the four strands engage in any single crossover event.

render gametes inviable (see discussion in Chapter 6, pp. 120–122). In addition to inviable meiotic products that reduce populational variability by reduced gamete diversity, physical restrictions may be imposed on crossing-over in chromosome regions that are heterozygous for the inversions.

FACTORS AFFECTING THE FERTILIZATION STAGE

Various factors influence recombination at the stage of zygote formation or fertilization, including the mode of dispersal of gametes or spores, reproductive barriers, habitat differences, and others which regulate the chances for gametes from different genotypes or populations to meet and fuse. The nature of the breeding or mating system may serve to illustrate the principles underlying such regulation of recombination. Self-fertilizing species usually are highly homozygous, and the zygotes that form from fusions of genetically identical or very similar gametes also are homozygous no matter how much crossing-over or reassortment of gentically identical chromosomes may occur. Homozygosity arises as a consequence of inbreeding; therefore, such a mating system provides a genetic damper on the generation of variability in the species. Cross-fertilizing species are more heterozygous than groups that inbreed, but regulation of the extent of variability produced by outbreeding may be a function of one or more features of the life style of the species. Such modifying factors as population size and structure, longevity of the individuals, and mode of dispersal of the zygotes or progeny, as well as other traits, all contribute to reduction or enhancement of the potential variability that may result from cross-fertilization. Asexually reproducing species, whether facultatively or obligately asexual, have undergone a variety of evolutionary modifications, and this has led to some balance between the constancy and flexibility components of the genetic system. Some examples of different breeding systems may help to illustrate the general principles.

Compromises Between Immediate Fitness and Flexibility

Heterozygosity and cross-fertilization constitute theoretically adaptive traits in species. The reserves of potential in such gene pools represent possible variations, which could prove advantageous under changing selection pressures when the predominant genotypes might become less adaptive in a new situation. Yet some species and groups of species have dispensed with such variability and remained successful. Many species of flowering plants have become self-fertilizing and thus homozygous, while others have become apomictic (substituting asexual for sexual reproduction) and sacrificed their variability. A better understanding of such evolutionary events comes from examination of specific features included in the life style of a species and the compensations that have been gained as a result of such reduction in potential variability.

Among the flowering plants documented so ably by G. Ledyard Stebbins and others, one or more of the following traits is characteristic of either apomictic or self-fertilizing descendant species with heterozygous cross-fertilizing ancestry. Such types often occur in temporary habitats, as is the case for the common dandelion, which is an asexually reproducing (apomictic) species. Lawns or open fields or disturbed areas such as roadsides abound with dandelions at certain times of the temperate growing season. These populations fluctuate rapidly and substantially in size. Such species also have an effective means for seed dispersal, as can be attested to by anyone who has observed the numerous fluffy gray seed-containing fruits produced shortly after the dandelion flowers appear on the lawn. Very often only one or a few vigorous immigrants in a new locality can produce a large population in just one or two generations. A similar syndrome of traits characterizes many self-fertilizing sexually reproducing flowering plants, such as the familiar thistles which quickly occupy roadsides, fields, and untended lawn or garden areas. Such species as the dandelion and thistle have accepted a high level of genetic constancy during their evolution and have gained advantages because of this. A well-adapted genotype in these species may produce many descendants, all with the same, or with equally well-suited, gene combinations as the parents. They become established very quickly and take over a habitat by virtue of large numbers of well-adapted progeny, which spread rapidly in a transient environment.

For those genetically constant species that produce an occasional new genotype—in a rare instance of outbreeding in a self-fertilizing species or of a sexual progeny for apomictic forms—it is possible that some variability becomes inserted into their gene pool. Such events lead to variation in the total gene pool of the species but to relatively little variability in any one population of a species. The members of a population may be genetically very similar or even identical but still be quite different from other populations of the same species. The production of occasional variants provides some flexibility for the species, but the principal components of the genetic system maintain the constancy of the population. The levels of compromise would depend on the frequency of events that introduce flexibility over the substantial background of genetic constancy of populations.

By contrast with species with a large component of immediate fitness, such as the dandelion, we would expect quite different adaptive features for the genetic system of cross-fertilizing species of flowering plants. Such populations would produce many descendants, but relatively few of these would survive to perpetuate the species. The species probably would be highly heterozygous and have a high index of recombination (higher chromosome number and more chiasmata per chromosome). They generally would be long-lived, would occupy relatively stable habitats, and would occur in populations of more or less constant size. An excellent example of such a flexible genetic system with high variability potential would be forest tree species such as the oaks. The adaptive features of a flexible genetic system in an oak species should be obvious from a consideration of its usual mode of existence. Forest populations undergo little short-term change; once established, the trees continue to occupy space and utilize the surroundings for many years. Progeny as fit as these successful parents, or more fit, are probably quite rare. But if numerous progeny with numerous genotypes are produced, then the probability is increased for some of these to be successful. Most of the progeny would be less fit than the parents and would not survive to perpetuate the species, but those few progeny with an appropriate genetic constitution would have a good chance of becoming established in the limited habitat space available. Flexibility predominates in such species.

Compromises between immediate fitness and flexibility do occur in highly heterozygous, cross-fertilizing flowering plants. Such species often are short-lived annual plants and occur in populations that undergo considerable fluctuation in size. There are many examples of such species among the flora of the western and southwestern United States, especially among members of the sunflower family (*Compositae*). Species growing in areas that receive considerable rainfall during different seasons or in different years but that are subjected to dry periods most of the time would tend to exist in populations that fluctuated in size according to the amount of water available. There would be advantages to an increasing degree of genetic constancy at the expense of flexibility in such outbreeding species. Constancy would be adaptive, since favorable gene combinations could persist for several generations under such stringent living conditions, but flexibility still could be achieved through the heterozygosity resulting from cross-fertilization. The evidence for compromise between immediate fitness and genetic flexibility can be observed in the chromosome complements of such species, as compared with related forms inhabiting friendlier surroundings.

Desert annuals usually have low recombination index traits, including fewer chromosomes and a lower chiasma frequency than their near relatives. The greater strength of genetic linkage preserves favorable gene combinations, but outcrossing generates diversity within and between populations. *Haplopappus gracilis* is a small desert annual species in the sunflower family which occurs in the southwestern United States. It has the lowest known chromosome number among the plants, just two pairs of chromosomes. At least one of these two chromosomes of the genome is believed to have been the result of centric fusion. The evidence in support of this interpretation consists in part of the observation that related species of *Haplopappus* contain four or eight chromosome pairs, which are much smaller than the very large chromosome of *H. gracilis*. Cytological study of chromosome homology during early prophase of meiosis in hybrids provided additional evidence of the suggested sequence of chromosome number decrease in this group of related species. Constancy has been strengthened because fewer recombinants are produced in *H. gracilis,* but enforced cross-fertilization permits heterozygosity and the attendant flexibility in the genetic system of this flowering plant.

From the extensive analysis of flowering plant genetic systems by various investigators we may make predictions about particular relations between growth habit, population structure, and the genetic system. Specifically, we would predict that long-lived perennials, especially those in stable habitats, would be characterized by the occurrence of sexual reproduction, cross-fertilization, relatively high chromosome number, and high crossover frequencies, all of which permit maximum recombination potential. If groups occupy temporary habitats and have reduced life-spans, then we would expect one of three different tendencies also to be present: (1) apomixis, that is, an asexual substitute for sexual reproduction; (2) self-fertilization; or (3) reduced index of recombination as established by reduced chromosome number and fewer chiasmata per genome.

ANIMAL GENETIC SYSTEMS

Among protists and invertebrates it is not unusual to find an emphasis on genetic constancy over flexibility, which would be predicted for species with short life-span and a population structure that fluctuates in size in short periods of time. In the ciliated protozoan *Paramecium* reproduction occurs most often by asexual fission, but other reproductive processes can occur which lead to increased genetic variability. Sporadic episodes of recombination are usually followed by relatively lengthy intervals of asexual multiplication and the continuation of genetic constancy that this makes possible. The restrictions on recombination in *Paramecium* are of a different magnitude from those we have reported for highly evolved flowering plants in terms of the chromosome complement, but they are similar in many ways to those we have described for apomictic flowering plants. Colonizing species of these types would be better adapted to their particular living conditions if constancy predominated the punctuating incidents leading to genetic variability.

Among those insects and fungi with alternate sexual and asexual phases in the life cycle, the compromise is such that genetic flexibility and constancy are present in different phases. The total genetic system provides an unusual level of advantage gained from the system's having both of these qualities. During the first part of the cycle when warmer weather begins, the sexual phase occurs. After a brief period of sexual reproduction during which variability is generated, the lengthier asexual phase is initiated. In the fungi, as in animals with similar life styles, the asexual generations occur in repeated rounds during the whole growing season. The spores produced during asexual reproduction in fungi are capable of vegetative growth and the production of great quantities of new asexual spores, and these bouts of reproduction continue for many generations during one season.

In insects such as the aphids, which also are known as "plant lice," the asexual process of parthenogenesis occurs during most of the growing season, after the brief sexual phase. In parthenogenesis only females are produced from unfertilized eggs. These asexual populations are produced rapidly and in large numbers, and they have the resulting high level of genetic constancy. There are many different populations, and these generally are initiated by many different genotypes of the species. During the sexual phase numerous progeny are produced, but only some of the genotypes prove to be successful in a particular habitat. Since a successful genotype may become established very quickly and produce tremendous numbers of new aphids by parthenogenesis, there is a high adaptive value to each aspect of the genetic system for such a species. Remarkably similar adaptive complexes occur in forms as diverse and unrelated as the fungi and insects, since each group has incorporated a set of requirements that ensure a more successful existence. Such similarities undoubtedly arose by completely different pathways of genetic change, resulting in analogous adaptations that are typical of convergent evolutionary phenomena. We will discuss such modification patterns in Chapter 10.

Sex as an Adaptive Complex

The pathway of evolution that led to a sexual mechanism of reproduction in eukaryotes is one of the unknowns in biology. Many coordinated processes participate in the accomplishment of a life cycle, and we know little about the origins of the sexual components of the cycle. With the fusion of nuclei to produce a zygote with twice the chromosome complement of either of the two parent cells, the compensatory mechanism of meiosis must be introduced to provide the means by which the chromosome number subsequently is reduced to the original level. Without meiosis, chromosome numbers would continue to double endlessly, which does not occur in existing forms and could not have led to success in the ancestral strains. The alternating sequence of fusion (fertilization is a particular kind of fusion in which a smaller gamete such as a sperm fuses with a larger gamete such as an egg cell to produce a zygote) and meiosis in a sexual life cycle introduces the alternating phases of diploidy, which is initiated by nuclear fusion, and haploidy, which is reinstated by meiosis in a diploid cell.

Simple life forms generally, but not always, have a sexual life cycle that characteristically is haploid for the longer interval and diploid only briefly. In such systems it is possible that the only diploid cell in the whole cycle is the zygote, and that meiosis occurs shortly or immediately after nuclear fusion. Meiosis restores the haploid condition, and the major portion of the species life-span is spent as the haploid individual. Although there are some variations, the sequence of life forms from protists to the fungi, plants, and animals reveals a consistent pattern of prolongation of the diploid phase and concomitant reduction in time for the haploid condition. The pattern is much clearer in plants than in either fungi or animals, but is detectable nevertheless in all these groups. Such sequences indicate that haploidy is the primeval condition and that diploidy evolved from it. Among the algae and fungi, which are simple non-tissue-forming organisms, every variation can be found, from species that are diploid only briefly as a zygote cell to others that are chiefly diploid and produce haploid cells that represent only a small portion of the life cycle. In some forms, such as the common yeast, either the haploid or the diploid condition may predominate, depending on conditions for growth or on the genetic program in different strains of the same species. Protists also show considerable variation. Ciliated protozoa such as *Paramecium* exist as diploid cells throughout their life; their particular method of nuclear distribution after meiosis results in the presence of pairs of haploid nuclei in each conjugating cell, which then fuse to restore the diploid nuclear state.

Mosses and liverworts are the simplest of the tissue-forming land plants, and it is the haploid phase individual that predominates and that we recognize immediately as the organism belonging to these groups. The diploid plant is also multicellular, but it is less conspicuous, shorter-lived than the haploid, and usually an outgrowth of the haploid plant. Those plant groups possessing vascular tissues that function in water and food conduction are the principal land flora. These include ferns and other pteridophytes, which do not produce seeds; the gymnosperms, such as pines, spruces, redwoods, cycads, and other groups, which bear naked seeds; and the angiosperms or flowering plants, which produce seeds enclosed in fruit (maternal) tissues. A clear progression shows the increasing predominance of the diploid phase; the haploid stage in ferns is represented by a tiny independent green plant that has a relatively brief existence, whereas in flowering plants haploid phases consist of only a few cells that exist for a very short time prior to internal fertilization of the egg by the sperm nucleus, which is carried to its destiny in the growing extension of the pollen grain.

The variable duration and prominence of the haploid phase typical of protists do not prevail among the multicellular animals. In animal species the haploid phase generally is represented only by the gamete products of meiosis. The gametes fuse almost immediately after formation and the diploid phase is rapidly restored. The diploid phase, obviously, is predominant.

The advantages of diploidy in the storage of variability and in preserving the organism from immediate selection against recessive alleles was discussed in Chapter 6 (pp. 126–127). Sexual reproduction is a means for establishing the diploid condition and therefore is one mechanism for generating genotype variability. Heterozygosity is not a condition of haploid cells, which contain only one chromosome of every kind and thus only one allele for every gene that is present in the complement. The diploid nucleus contains two chromosomes of every kind and thus provides the basic requirement for developing heterozygosity, since different alleles may reside in the homologous chromosome pairs. The degree of heterozygosity that may develop in sexual forms would depend on the relative merits of immediate fitness and flexibility under the direction of natural selection.

EVOLUTION OF SEX-DETERMINATION MECHANISMS

Cell or nuclear fusions that are followed by meiosis in one life cycle constitute sexual acts. The fertilization → meiosis → fertilization sequence in which haploidy and diploidy alternate in one generation serves as a standard by which sexual reproduction may be recognized, even in unusual variations. Whether the fusing components are morphologically identical or not, many species possess a mechanism of sex determination, since genetically identical cells usually cannot initiate the new generation. The simplest system of sex determination is based on a single pair of alleles, each allele of the gene determining a genetic population that is compatible only with members of a population having the alternative allele (Fig. 9.2). Genetic analysis reveals the monohybrid inheritance pattern, and experimental evidence clearly shows that individuals that are genetically of one mating type are incompatible with others of the same mating type but compatible with the alternative genetic group. Several conventions are used to designate mating type systems of this simple sort, but generally either *plus* and *minus,* or *A* and *a,* or *a* and α serve to identify the alternative alleles or the two mating types of the system. More than one pair of alleles may underlie the mating type system, and more than two mating types may exist in some groups, including the protozoa and the fungi. The patterns are comparable however, a system of one or a few genes which assort and determine the sexual competency of the individual to mate with another.

SEXUAL DIMORPHISM There is substantial variation among simple non-tissue-forming eukaryotes as to cells that are competent to initiate the new generation subsequent to nuclear fusion. Many protists have no specialized gamete cells, and fusions may occur between morphologically identical units, which may also function as the vegetative components of the system. Paramecia conjugate and the ex-conjugants undergo subsequent fissions, which give rise to the new generations of cells that are morphologically indistinguishable from each other and from the initial conjugants. Multicellular algae and fungi provide numerous examples of sexual events in which fusions may occur between identical components or cells, which can be recognized clearly as gametes that have no other function in the

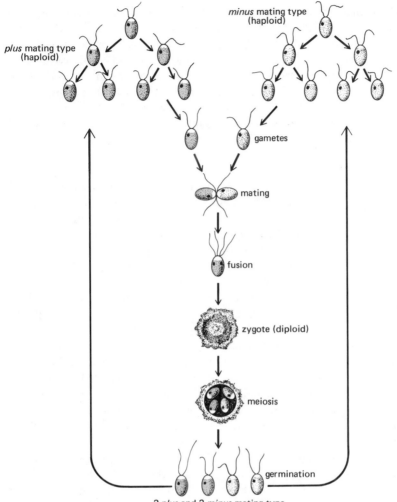

Figure 9.2   Sex determination in the unicellular green flagellate *Chlamydomonas*. Haploid cells of either mating type may multiply vegetatively by mitosis, but fusions which initiate the sexual phase of the life cycle can occur only between gametes that are of different mating types. A single gene existing as *plus* and *minus* alleles underlies sexuality in this species.

life cycle than sexual fusion. Gametes may be produced in specialized structures or organs which differentiate from the vegetative body, or such structures may be lacking altogether. Every kind of variation and intermediate system of gamete production and morphology seems to have evolved in these groups, in addition to the extremes mentioned above.

There are multiple pathways among the phylogenetically simpler organisms, but the generality that emerges is that species having similar gametes diverged to produce forms in which the gametes differed genetically, biochemically, and morphologically from each other. The behavioral differences between gametes usually are expressed in the migration of one gamete to the location of the other kind of sex cell. Among the metazoans and in the complex plants of the mosses, liverworts, and fern groups, the motile smaller gamete is the male sexual product, while the nonmotile larger egg cell is produced by the female sexual component. Gymnosperms and angiosperms, representing the most highly evolved plant forms, produce pollen which differentiates to produce sperm nuclei that travel within the pollen vehicle to reach the nonmotile egg cell. The male gamete is the mobile element in all these cases, although the manifestations of mobility may vary. The larger size of the egg cell usually is a reflection of the stored nutrients, which are contained along with protein synthesizing systems that serve to provide the energy and chemical raw materials utilized by the embryonic representation of the next generation that develops from the zygote (or fertilized egg). The size difference between sperm and egg is especially marked in birds, which produce eggs that are many million times larger than the minute spermatozoa. In addition to size and mobility, male gametes generally are produced in multitudes, compared with the numbers of egg cells produced. This particular distinction undoubtedly reflects an adaptive feature of the more hazardous existence of the male sex cells, providing greater insurance that some of the many sperm will be successful in their journey to the egg.

The development of dissimilar gametes almost certainly preceded the evolution of sexually dimorphic patterns in which different individuals were exclusively of one sex or the other. We would expect the earlier forms to have been hermaphroditic, having the capacity to produce both kinds of gametes in a single individual. Many of the simpler animal groups are primarily hermaphroditic, but all the highly evolved animals characteristically produce individuals having only one sex capacity, males and females, producing sperm or eggs, respectively. Among the plants the predominant pattern is hermaphroditism, but the terminology generally used is different. The flowering plants principally produce male and female structures in the same flower, these being called perfect flowers. Species that produce both male flowers and female flowers (only one kind of sex organ in each) on the same plant are called monoecious; if male and female flowers are produced on different individuals, the sexual pattern is dioecious. Elaborate mechanisms to ensure cross-pollination are known for flowering plants so that many of these are not functional hermaphrodites even if they produce perfect flowers or are monoecious. Among hermaphroditic invertebrate animals the frequency of self-fertilization usually is low; instead, cross-fertilization may occur between different individuals, which accept sperm that fertilize their eggs at the same time they release sperm to fertilize the eggs of their partners. Whether there are separate sexes, as in most animals, or mechanisms that ensure crossbreeding between different hermaphroditic individuals, as in many plants, these inventions lead to heterozygosity and to the continuing diversity of the species gene pool.

It is generally accepted that dioecious angiosperms evolved from monoecious ancestry. Some species groups, such as the maples, contain both of these patterns; the genetic differentiation does not appear to be particularly great if they occur in the same genus. There is one interesting case of genetic variation in corn (*Zea mays*), which sometimes has been invoked as a model to explain the origin of dioecious lines from monoecious predecessors. In corn of the usual *FFMM* genotype the pollen is produced in the tassel flowers at the apex of the plant, and eggs are produced in the flowers of the "ear" on the stem of the same plant. A mutant allele (*f*) is known to produce female-sterile individuals, the homozygous (*ff*) plants producing viable gametes only in the male flowers of the tassel. Another mutant allele (*m*) leads to exclusively female plants in which the flowers of both the tassel and the ear produce only eggs and no pollen. It is possible to perform crosses to construct exclusively male individuals of the genotype *ffMM* and exclusively females of the genotype *ffmm*. If such populations were to develop and stabilize so that the predominant genotypes were *ffmm* (female) and *ffMm* (male), then a permanently dioecious group might be perpetuated, which would continue to produce these two genotypes every generation. Considering the apparently simple genetic differentiation of monoecious and dioecious species, which belong to the same genus in some cases, it would not be unreasonable to propose that dioecism did originate from such minimal genetic modifications. If there were adaptive value to this pattern of sexual differentiation, then modifying mutations that fixed this system in the population would be advantageous and would be more likely to become incorporated in the populations. In this way a more substantial genetic foundation would be established for the new sex pattern in the descendant lineages.

SEX CHROMOSOMES IN ANIMALS   The commonest pattern of sex determination in animals is based on sex chromosomes rather than on one or a few genes (Fig. 9.3). A sex chromosome system is more likely to perpetuate the sexes, in equal frequencies for most species, and not be readily exposed to modifications by one or a few mutational events. The fundamental XX-XY chromosome pattern is present in most animals although the sex-determining roles of the X and the Y chromosomes appear to be different, at least in the insects and the mammals. Sex determination in *Drosophila* has been studied intensively and served until quite recently as a general model for all species with XX females and XY males, including mammals.

In *Drosophila* the sexes are determined by the balance between the num-

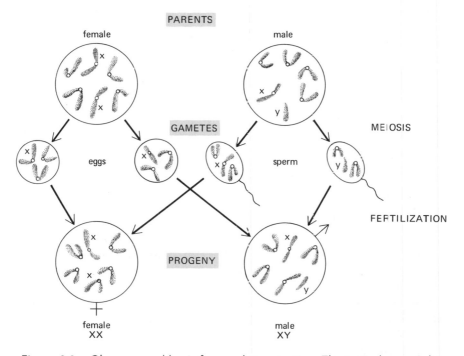

Figure 9.3   Chromosomal basis for sex determination. The typical mammalian scheme involves XX females, which produce one kind of egg cell (all have one of the X chromosomes), and XY males, which produce X-bearing or Y-bearing motile sperm in approximately equal frequency. The sex of the progeny depends on the particular kind of sperm that fertilizes the egg.

ber of autosome sets and the number of X chromosomes in the adult cells. In the company of the conventional two sets of autosomes of the diploid species, females resulted when two X chromosomes also were present, whereas males contained either the usual X and Y chromosomes or only a single X chromosome and no Y. Converting these observations to a general scheme, it was proposed that the ratio of the number of X chromosomes to autosome sets determine sex—males when it was 0.5 (X : AA) and females when the ratio was 1.0 (XX : AA). Thus in the usual diploid populations, males could either be XY or XO (lacking the Y), whereas females were XX. According to this hypothesis, ratios that varied from 0.5 for males and 1.0 for females would lead to sexual aberrations. Such aberrant forms were found (Table 9.1), and they substantiated the predictions made by the model so well that the generalized Balance Theory of Sex was proposed. The postulates of this theory were assumed to apply to all species with an XX-XY mechanism of sex determination. The Y chromo-

TABLE 9.1 SEX DETERMINATION IS BASED UPON A BALANCE BETWEEN
NUMBER OF AUTOSOME SETS AND X CHROMOSOMES IN
THE DIPLOID ADULTS OF *DROSOPHILA*

Sex chromosomes present	Number of sets of autosomes	Ratio of X : A	Sex of the individual
XX	2	1.00	female
XY	2	0.50	male
XXX	2	1.50	metafemale (sterile)
XXXX	3	1.33	metafemale (sterile)
XXY	2	1.00	female
XXX	4	0.75	intersex
XX	3	0.67	intersex
XY	3	0.33	metamale (sterile)
X	2	0.50	male (sterile)

some was dispensable, although XO males in *Drosophila* were sterile, in part because of the absence of the genes for sperm motility, which are situated in this chromosome. Supporting evidence for the general tenets of the Balance Theory was provided by the many studies of chromosome patterns in various other insect groups, but especially among the grasshopper species of the order Orthoptera (*Drosophila* belongs to the Diptera, the order of insects that includes all the two-winged insects). Most species of Orthoptera have XX females and XO males, so the X : A ratio is 0.5 for males (XO/AA) and 1.0 for females (XX/AA).

Recent studies of sex chromosomes and sex-determination patterns in humans and other mammals have shown that a different system operates from the one described in the Balance Theory. In humans, who have been studied in great detail since the development of suitable methods for human chromosome preparations from cell cultures in 1956, the Y chromosome is strongly male-determining (Table 9.2). Females may have the usual two X chromosomes or one or three or more, along with the normal 22 pairs of autosomes. Therefore, in humans and other mammals studied, XO leads to female differentiation and not to the development of males, as in *Drosophila* and grasshoppers. Human females with a single X chromosome develop the clinical symptoms of Turner's syndrome. In the mouse, however, there are no aberrant effects, whether the females develop under the guidance of an XO chromosome constitution or the normal XX condition. Human males have 46 chromosomes, the 44 autosomes as in females, but an XY sex chromosome pair instead of XX (see Fig. 6.11). Clinical cytogenetic studies have shown that a relatively high frequency of human males have 47 chromosomes, the extra one being an additional X chromosome. There may be as many as one in every 400 males born with an XXY chromosome constitution, which leads to the clinical symptoms known as Klinefelter's syndrome. These individuals are morphologically and anatomically males, but they possess a number of feminized traits and often show some degree of mental retardation, as well as being sterile. Variations more extreme than XXY have been found, including males with Klinefelter's syndrome who have as many as four X chromosomes along with 44 autosomes and a Y chromosome. Males with XXYY + 44 autosomes also have this clinical syndrome, so what is required for normal development is not simply a matter of the proportion between X and Y chromosomes.

Many of us have seen calico cats whose fur includes both yellow and black coloration. These fur colors are expressions of a pair of sex-linked alleles for one gene; females may have one allele on each chromosome of the pair, but males have only one X chromosome, so they have either the allele for black or the allele for yellow fur color. It follows then that calico cats must be females; since the two colors are produced, two alleles and two X chromosomes must be present. Occasional calico males have been studied, and in each case it has proven to be an XXY male that develops these particular fur color patterns. Sex chromosome aberrations obviously are not the private property of humans among the mammalian species, nor would one expect them to be.

A number of cases have been described recently of male with 47 chromosomes, the extra chromosome being another Y. There is insufficient evidence at present to know whether the observed "antisocial" traits of such XYY males is a genetic consequence of the chromosomal constitution or an outcome of psychological trauma resulting from the differences between the adolescent development of such individuals and that of the usual XY males. Males who are XYY usually grow at a greater rate than XY boys of comparable age and usually score lower on general tests for intelligence than is the norm. Such peer group differences may become sources of conflicts, which could be expressed in adulthood as antisocial tendencies. Further careful analysis of males from the general population are required, in

TABLE 9.2   SEX DETERMINATION AND SEX CHROMOSOME ANOMALIES IN HUMANS

Individual designation	Chromosome constitution	Total number of chromosomes	Sex	Fertility
Normal male	AA XY	46	male	+
Normal female	AA XX	46	female	+
Klinefelter types	AA XXY	47	male	−
	AA XXXY	48	male	−
	AA XXXXY	49	male	−
	AA XXYY	48	male	?
	AA XXXYY	49	male	?
Turner syndrome	AA X	45	female	−
Triplo-X	AA XXX	47	female	±
Tetra-X	AA XXXX	48	female	?
Penta-X	AA XXXXX	49	female	?
XYY-male	AA XYY	47	male	+

addition to the relatively large samplings of males in penal institutions and similar places.

There are variations on the basic theme of XX-XY sex determination in animals (Table 9.3). Birds, reptiles, some fishes, and insects such as those in the butterfly group (order Lepidoptera), among others, are characterized by homogametic males and heterogametic females. In these species it is the female that produces two kinds of eggs and the male that produces only one chromosomal type of sperm. The convention has been to use different letters to symbolize such systems, the most frequent scheme is one in which females are WZ and males are ZZ. We will not confuse the issue with these symbols, but will refer instead to XY females and XX males in such species.

A wealth of diversity in sex chromosome constitutions has been described for different insect groups, in addition to the commoner patterns of XX-XY and XX-XO. Among the mantids and grasshoppers such variations on the XX-XO scheme have been described as species with XXXX females and XXY males, with genetic differences that distinguish one of the X-chromosomes from the other so that we actually find $X_1X_1X_2X_2$ females and X_1X_2Y males. Pairing between homologous X chromosomes during meiosis leads to a high degree of normal segregation of the different X chromosomes during gamete formation and to the retention of a standard sex chromosome complement in the adults that develop subsequent to gamete fusions.

Very few plants have been found to have sex chromosomes, but in those flowering plants that do have such a genome component it is the diploid pollen-producing males that are XY and the diploid egg-producing females that are XX. The simpler mosses and liverworts have sex chromosomes, and the brief diploid phase is XY. Upon the formation of haploid spores at meiosis in diploid cells, the spores are released and develop under suitable conditions into the longer-lived haploid gametophytes (gamete-producing plants). Haploid plants with an X chromosome are female, and those gametophytes with a Y chromosome are the male plants which produce motile sperm. Under normal conditions there are no diploids other than the XY type; the fusion of egg (X) and sperm (Y) routinely leads to the sexually neutral diploid sporophyte (spore-producing plant).

TABLE 9.3   SUMMARY OF MAJOR CHROMOSOMAL SEX DETERMINATION PATTERNS

Males	Females	Species in which these occur
XY	XX	mammals, *Drosophila*, some flowering plants
XO	XX	most grasshoppers, many *Hemiptera* (bees, wasps, ants)
XX	XY	some fishes, reptiles, birds, Lepidoptera (butterfles and moths)
Y	X	liverworts

SEX DIFFERENTIATION   The determination of sex of the individual is achieved at the moment of fertilization, depending on the combination of sex chromosomes in the gamete fusion product that will develop into the new individual (see Fig. 9.3). The subsequent differentiation of sexual expression in the organism is quite complex and depends on the activities of numerous genes in the whole complement of chromosomes, as well as on the local conditions within the individual and in its external environment in which such gene action takes place. An often-quoted example of environmentally-determined sex is the marine worm *Bonnelia*. The larva is a small free-swimming form, which eventually settles to the bottom of the sea where it undergoes further development. If the larva happens to fall on the proboscis of a female, it will enter her body and differentiate into a minute male and lead a parasitic existence there. If the larva settles on the sea bottom, it will differentiate into a female, at least 500 times larger than the male of the species. By rearing larvae in seawater to which an extract of female proboscis tissue has been added, it can be shown very clearly that sex determination is environmentally controlled, since all the larvae will differentiate into males. Lacking such an extract, females will differentiate from the sexually undetermined larvae.

Sexual differentiation cannot be under exclusive genetic control in hermaphroditic forms; individuals of a single genotype produce both kinds of sex organs, as in most flowering plants and in such invertebrates as the earthworm. In those diploid species in which there is a separation of the sexes, the determination of sex differentiation has been brought under greater genetic control during evolution. Similarly, simpler genetic control systems can operate in predominantly haploid organisms, since the diploid phase is sexually neutral and the haploid segregants may be of functionally different sexes, whether or not they are otherwise distinguishable.

An interesting type of sex determination mechanism has been described for the Hymenoptera, the order of insects that includes bees, wasps, and ants. In these species the female develops from diploid fertilized eggs; males develop from haploid eggs which have not been fertilized. The male contributes his total chromosome complement to his daughters, whereas the males themselves have no fathers and can produce no sons. Male hymenopterans produce genetically identical sperm as a consequence of a meiotic process during which one of the two divisions is aborted so that the other meiotic division results in sperm that have an intact haploid genome like that of the adult male. In the honey bee two kinds of females differentiate. All are diploid, having hatched from fertilized eggs, but workers ordinarily do not reproduce whereas the whole function of the queen bee is to lay the eggs through which the colony will be perpetuated. The factor that leads to differences in appearance and behavior of these genetically equivalent female types apparently is the diet provided to them while in the larval stage. The royal jelly, fed to those females destined to become queens, is rich in the vitamin pantothenic acid; the worker larvae receive a different diet. There is a genetic basis for sex determination in such species, but the differentiation into sexually functional individuals can be modified by subsequent environmental conditions. The social bee (*Melipora*) produces diploid females that differentiate as sterile workers or as sexually functional queens depending (unlike the honey bee) on the degree of genetic heterozygosity of the larvae. The queens develop from among the most highly heterozygous female larvae whereas less heterozygous diploid larvae become the workers of the colony. Diet apparently is not a factor in the differentiation of sex expression for females in this group.

A somewhat different sex-determination mechanism has been described for some kinds of wasps in which wild populations ordinarily contain diploid females and haploid males. The basic difference would appear to be based on the ploidy level, but experimental studies have revealed that diploid larvae can develop into males. The crucial distinction between the two sexes is the level of heterozygosity versus homozygosity, and not ploidy. Since haploids have only one set of chromosomes they cannot be heterozygous, there being only one allele for each gene in the complement. Haploids therefore develop into males among these wasps. Because the species are outbreeding, and homozygosity is rare or absent, diploids develop into females. But when homozygous diploids are produced in the laboratory, they invariably differentiate as males. The sex-determining mechanism is one of heterozygosity versus homozygosity; the production of fertilized eggs, which are genetically heterozygous, ensures the development of females, while haploid unfertilized eggs provide the only means for producing homozygous male progeny in natural populations. The term homozygous is a partial misnomer in the case of haploids in that there is only one genome present. The correct term is hemizygous, but for our purposes this is merely a detail of terminology.

Sex reversal phenomena provide interesting insights into mechanisms of sex differentiation. Frogs and toads apparently are easily subject to such reversal when the temperature is raised to about 32°C in their surroundings. At this temperature a genetically female amphibian will develop into a fertile male upon the maturation of the larval tadpole to the adult form. When normal females are mated with sex-reversed males, only female offspring will be produced because the two parent animals have identical sex chromosome complements. This example indicates that the chromosomal mechanism remains unchanged relative to subsequent generations, but that individual animals may respond to some conditions that lead to the inhibition of normal sex differentiation and may even result in differentiation into an individual of the opposite sex. Another peculiar example of environmental conditioning that overrides the sex chromosome mechanism and conventional sexual differentiation is found in cases where normal hens undergo sex reversal and become functional roosters that can father chicks. The normal hen has only one ovary, which differentiates into a functional egg-producing organ; the other gonad (sex organ) remains undifferentiated. If the functional ovary is destroyed, the other gonad may undergo differentiation of the sex cords to become a functional testis. Changes in the behavior and appearance of the sex-reversed chicken are due to the production of male sex hormone by the newly differentiated testis; both the primary and the secondary sex traits are altered when the testis develops. In humans, effects of sex hormones often lead to modifications of secondary sex characteristics, such as hairiness, pitch of the voice, breast development, and other features. But sex reversal at the level of gonad function is unknown for mammals.

Sex hormone influences can be seen in vertebrates in which patterns of cellular differentiation are dependent on such external influences as the circulating hormonal products of the endocrine glands. One such example is the freemartin in cattle. When twins of opposite sex are born, the freemartin female calf develops into a sterile animal as a result of modifications in its gonads under the influence of circulating male hormones produced by the male co-twin before birth. The hormone system leading to male differentiation in cattle becomes active earlier than the female differentiating hormonal system. The twins share common circulation through the placenta so that circulating hormones, as well as nutrients and oxygen in the blood, are accessible to both siblings. The freemartin never becomes a functional male, although it does differentiate as an intersexual animal as a result of its gonadal hormone production.

The situation is quite different in insects because their cellular differentiation is autonomous. Each cell develops according to its genetic potential; thus mosaics may be produced if the individual cells have different genes or chromosomes. Such cellular differentiation may arise from somatic mutations, which affect only the original mutant cell and those descended from it during development, or from aberrations in chromosome distribution during mitosis as the multicellular animal develops. Where the aberration involves a sex chromosome, some parts of the insect are female and other parts are totally male. Extreme examples of sex mosaicism in insects include the occasional gynandromorphs of *Drosophila,* which are half male and half female along a sagittal plane. Here, apparently, the first mitotic division of the fertilized egg was aberrant, so that in the two-cell stage one of the daughter cells retained the initial XX chromosome constitution and the other daughter cell lost an X chromosome. The cells derived from the XX cell initial are female and those from the XO cell initial differentiate as typically male tissues.

In these selected examples of sex determination and sexual differentiation the underlying theme has been that any cell has the basic potential to develop either male or female characteristics, rather than that there is an irrevocable transformation of the individual into one sex or another. It is the reaction system in operation which ultimately leads to the genotypic and phenotypic sex of the organism. Basically one mode of operation of the system directs the development of a male individual, and an alternate operation of the system leads to a female. The alternations of the system may be entirely environmental, as in the *Bonnelia* example, or genic, as in the simple mechanism in some protists; or, in many animals, they may depend on the sex chromosome mechanism. Sexual differentiation itself does not necessarily follow an undeviating pathway to a particular end result, inasmuch as modulating influences may affect the expression of the potential. Some of these modulating influences, such as temperature or nutrition, lead to the formation of intersexes or various aberrancies, or even to functional sex reversal. Relatively few inherited traits remain entirely unaffected by genetic and environmental influences during the development of gene expression. The genetic components of the sex-determining and sex-differentiating systems thus are not unusual in bowing to the effects of factors external to themselves.

SOME REFINEMENTS IN THE SYSTEM   The sex ratio at birth approaches 1 : 1 in those genetic systems based on a single pair of alleles or on two kinds of differentiated sex chromosomes, such as the X and Y in animals (Fig. 9.4). Variations in this ratio are realized by some systems that are based on genetic factors, but they occur more often in systems in which the genetic mechanism does not influence sex determination or differentiation. Insects such as *Drosophila* or the grasshopper will produce equal frequencies of male and females, based on the distribution of sex chromosomes to the gametes at meiosis and their combinations at fertilization. Other insects, such as the hymenopterans, may produce variable sex ratios, depending on the numbers of fertilized and unfertilized eggs produced. But the overabundant production of females in these bees, wasps, and ants has no direct function in the reproductive pattern of the population. Most of the fertilized eggs are destined to become the sterile workers that the colony depends on for its daily existence. Only an occasional fertilized egg differentiates as the functional queen of the colony, but this individual has the profound responsibility of perpetuating the population and the species. The actual ratio of functional females to males probably approaches the same 1 : 1 ratio as that in species with more conventional mechanisms for modulating the reproductive potential and the perpetuation of the species. Usually more males than females are produced, but a higher mortality rate operates among the males. There is a higher mortality for males in vertebrate animals, too, but the sex ratio at birth probably is approximately 1 : 1.

The separation of the sexes into male and female individuals provides better insurance for maximizing the genetic recombination potential since cross-fertilization is mandatory to perpetuate the species. The security of

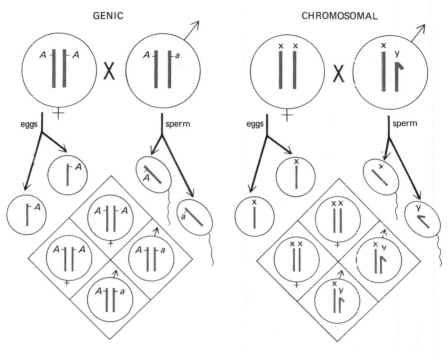

GENIC CHROMOSOMAL

1 FEMALE : 1 MALE   1 FEMALE : 1 MALE

Figure 9.4   Sex determination. The production of males and females in a
1 : 1 ratio can occur whether the sex determination mechanism is premised on
a pair of alleles of one gene or on a pair of differentiated sex chromosomes.
In each case one sex is homozygous and the other is heterozygous, whether
for an allele or a kind of sex chromosome.

this system is strengthened if restrictions are imposed on exchanges between
the sex chromosome types. If there were free crossing-over between the
X and Y chromosomes, there would be disruptions of the intactness of
these components, which could prove to be inadaptive. If crossing-over
between the different sex chromosomes is prevented or restricted, then all
or most of the X and Y chromosomes would be transmitted to the gametes
with the greatest retention of accurate information content. Gamete fusions
thus would lead to zygotes with whole complements of sex chromosome
genes and to less ambiguous sexual differentiation.

During evolution there must have been genic changes that diminished or
prevented crossing-over between the X and Y chromosomes but had little
or no effect on crossing-over frequencies between homologous autosomes in
the same nucleus. In some species there is no crossing-over between any of
the chromosomes in the meiotic nucleus of the heterogametic sex, although

crossing-over remains unaffected in the homogametic sex. Crossing-over thus does not occur in the male germ line of *Drosophila* or in the female germ line of the silkworm (*Bombyx*). Each of these is the heterogametic sex of the species. In those animal systems that have been studied extensively, especially among the insects and the vertebrates, there often is some homologous segment between the X and the Y chromosomes, and a low frequency of recombination can be detected by genetic methods. Most of the evidence for homology and crossing-over has been based on cytological rather than genetic evidence; therefore, definitive conclusions concerning sex chromosome recombinants cannot be made for any species that has not been the subject of breeding or other genetic analysis. In these species the principal segments of the sex chromosomes do not undergo exchanges, however, so that the bulk of the genetic information remains intact when parceled out to the resulting gametes at the end of the meiotic division sequences.

Some Suggested Readings

Crew, F. A. E., *Sex-Determination.* 4th ed. New York, Dover, 1965.

Darlington, C. D., *The Evolution of Genetic Systems.* 2nd ed. New York, Basic Books, 1958.

Robertson, D. R., "Social control of sex reversal in a coral-reef fish." *Science* 177 (1972), 1007.

Smith, H. (ed.), *Evolution of Genetic Systems.* Brookhaven Symposia in Biology, Number 23. New York, Gordon and Breach, 1972.

Stebbins, G. L., *Chromosomal Evolution in Higher Plants.* Reading, Massachusetts, Addison-Wesley, 1971, pp. 72–123.

PART III

TRENDS IN EVOLUTION

CHAPTER 10

PATTERNS
OF
EVOLUTION

The geological and fossil records constitute the single most powerful line of evidence for evolution as a process with a time dimension. Yet, even if there were no trace of former life we still would feel confident that such a progression had occurred, because the evidence from comparative anatomy, morphology, physiology, biochemistry, and genetics clearly show relationships within and between groups of organisms. The evidence from comparative studies of living species indicates a nonrandom distribution of genetically determined traits. Support for descent with modification rather than random combinations of traits comes from the numerous observations of greater resemblance between genetically related species than between distantly related ones. If such similarities occur as constant patterns, the most reasonable interpretation would seem to be that such species are phylogenetically related. Inasmuch as living species exhibit such genetic relationship, then we would predict that the fossil record also should show such patterns of relationship. Sets of similarities not only should typify those species that have been preserved in the fossil record, but also should be traceable through time to modern forms to comprise phylogenetic sequences that extend across the time dimension from the ancient past to the present day. Such consistencies in relationships do exist, and the fossil record thus provides strong support for the Darwinian principle of descent with modification leading to groups of modern species. To appreciate the fossil record more fully we should know some basic characteristics of the time dimension as it has been established for the underlying geology with which the fossil record has been inextricably entwined.

Some Geological
Features

The divisions of geological time are somewhat arbitrary, and new evidence can, and does, lead to some modification of the absolute time intervals previously assigned to any of these divisions. For example, the Paleozoic era is recognized as beginning with the first fossil members of all the modern animal phyla, including chordates (Table 10.1). The ear-

TABLE 10.1 SUMMARY OF MAJOR HIGHLIGHTS OF THE GEOLOGICAL AND FOSSIL RECORDS

Eras*	Periods*	Epochs	Geological features	Aquatic life	Terrestrial life
Cenozoic 63 ± 2	Quaternary 3 ± 2	Recent	Warmer climates	All modern groups present	Modern humans
		Pleistocene	Periodic glaciation		First hominines
	Tertiary 63 ± 2	Pliocene	Mountains rising		Large mammals; hominids and pongids
		Miocene	Cooler climates		Hominoids
		Oligocene	Warmer: lands low		Anthropoids; many birds
		Eocene	Few inland seas		Modern mammals; herbaceous flowering plants
		Paleocene			Woody plants; birds; mammals
Mesozoic 230 ± 10	Cretaceous 135 ± 5		Mountain building; inland seas and swamps	Modern bony fishes abundant; extinction of aquatic reptiles	Extinction of dinosaurs; first modern birds; rise of woody plants; archaic mammals
	Jurassic 180 ± 5		Continents high; seas shallow	Aquatic reptiles, skates, rays, and bony fish abundant	Dinosaurs dominant; first birds, mammals, flowering plants; insects and conifers abundant
	Triassic 230 ± 10		Warm climates; many deserts	First aquatic reptiles; rise of bony fishes	Adaptive radiation of reptiles; seed ferns extinct
Paleozoic 600 ± 50	Permian 280 ± 10		Mountain building; glaciers; aridity	Widespread extinction (trilobites, placoderms, etc.)	Widespread extinction; reptiles abundant; therapsids; insects
	Carboniferous 345 ± 10	Pennsylvanian	Warm, humid climate; coal swamps	Bony fishes	First reptiles; insects common; many land plants
		Mississippian	Warm and humid; then cooler	Abundant sharks	Amphibians abundant; first insects; forests

Period	Physical conditions	Life (marine)	Life (land)
Devonian 405 ± 10	Periodic glaciation; more aridity	Sharks and fishes abundant; many invertebrates	First amphibians; spiders, forests; first gymnosperms
Silurian 425 ± 10	Inland seas	Ostracoderms; many algae	First land plants; land invertebrates
Ordovician 500 ± 10	Mild climate; inland seas	Invertebrates abundant; first vertebrates	none
Cambrian 600 ± 50	Mild climates; inland seas	Trilobites common; many invertebrates; first chordates (tunicates); algae	none
Precambrian 4600	Glaciation; volcanic activity; mountain building	Fossils rare; prokaryotes predominant; first eukaryotes appeared	none

* Numbers indicate approximate times the eras and periods began, in millions of years.

liest subdivision of the Paleozoic era is called the Cambrian period, and a new discovery of some Cambrian type of fossil in rock formations older than had been known before would be one basis for putting the beginning of the era back to an earlier date. Such discoveries have been made, and the present date for the initiation of the Paleozoic era is considered to be 600 million years ago (plus or minus 50 million years). Because we cannot be absolutely sure of the precise date the statistical variation of *plus or minus* 50 million years designates the range of uncertainty around the mean (average) value. Thus the Cambrian period probably began some time between 550 and 650 million years ago. The mean value of 600 million years continues to be used for convenience and approximate accuracy based on the available evidence.

The time divisions essentially are established in correspondence with gaps or discontinuities in the geological or fossil record. The more drastic the change in the record (the larger the discontinuity), the greater is the magnitude of the designated time interval. The largest time divisions are *eras,* which are based principally on major mountain-building episodes. Era duration has also been based in part on the nature of the fossil record. The Paleozoic era, therefore, is delineated on two criteria: mountain-building episodes and the life forms preserved in the fossil record of the formations laid down between two major geological episodes. The first complex life forms are Paleozoic fossils. Five eras have been recognized by geologists, but more recent usage from a biological point of view has led to the general designation of the first four billion years of the Earth's history as the pre-Cambrian era rather than as the Archaeozoic and Proterozoic eras. Major mountain-building episodes occurred, and the fragmentary remains of those uplifts have been recognized in different parts of the world. The Laurentian Mountains in Canada represent a part of those ancient pre-Cambrian uplifts. Fossils dated as 3 billion years old have been found in such persisting formations in various parts of the world. Such ancient evidence is rare, but there is excellent evidence of a progression of forms from the simpler prokaryotes in the oldest formations to more sophisticated prokaryotes and simple eukaryotes in progressively younger pre-Cambrian strata. The three eras that encompass the most recent 600 million years of planetary history are very rich in fossil deposits and better defined than the 4 billion years of the pre-Cambrian era (Fig. 10.1).

The mountain-building episodes included the upthrusts of the Appalachian Mountains in the United States and the Ural Mountains in the Eurasian region of the Soviet Union, which marked the end of the Paleozoic era; the events that produced the Rocky Mountains of North America and the Andes Mountains of South America, essentially a pole-to-pole upheaval, which marked the end of the Mesozoic era; and the major geological episode that occurred just prior to the Pleistocene epoch of the Cenozoic era and led to the Alps, the Himalayas, and the final upthrusts of

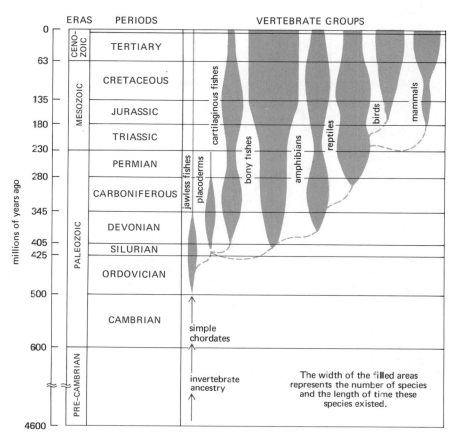

Figure 10.1   Evolutionary history of the vertebrates.

the Rocky Mountain range. These mountain-building incidents may have occupied millions of years, and almost certainly led to drastic alterations in those fossils included in the affected geological formations. Such upheaval almost certainly is responsible for the poor fossil record of those times and for the obliteration of fossils that had been deposited during earlier times in those regions of the world.

Beginning with the Paleozoic era the fossil record is sufficiently rich to designate subdivisions of the eras, based on both the fossil progressions and geological events of lesser magnitude than mountain building. The first of the *periods* of the Paleozoic era is the Cambrian, marked by the presence for the first time of representatives of all the modern phyla of animals. One invertebrate fossil type that serves as a useful marker of Cambrian strata is the trilobite (Fig. 10.2), a marine arthropod distantly related to crustaceans. Trilobites were abundant and diverse during the Cambrian period and many groups of these animals disappeared toward its end, the last trilo-

Figure 10.2 A fossil specimen of a Cambrian trilobite. (Photograph courtesy of the American Museum of Natural History, New York.)

bites becoming extinct by the end of the era itself. The beginning of the Ordovician period generally can be determined by its unique fossil record in conjunction with the absence of many specific groups of trilobites. The Permian period was about 50 million years in duration, and ended the Paleozoic era about 230 million years ago (plus or minus 10 million years). There are three periods of the Mesozoic era, which lasted for a total of about 170 million years during which a variety of evolutionary changes produced new life forms never found before in the fossil record. The periods of the Cenozoic era have been subdivided still further, into *epochs,* because such detailed information is available that finer guidelines can be discerned in the historical record of this most recent era. In addition to this, the Cenozoic era is the Age of Mammals (as the Mesozoic era was the Age of Reptiles), and our own history is buried in these fossil deposits. Our anthropomorphic curiosity leads us to seek finer guidelines for the reconstruction of the events that produced the highest forms of life, our own among them, in the Cenozoic era.

Fossils are found predominantly in *sedimentary* rocks and usually are absent in formations of *igneous* or *metamorphic* origin. Igneous rocks are volcanic in origin or form part of the original crust of the Earth, basalt and granite being typical examples of this group. One would not expect to find fossils in any formation deposited before life originated; thus they would be absent from crustal rocks. Nor would fossils be preserved in volcanic rocks formed deep within the Earth under conditions of very high pressure and temperature. There are, however, special conditions under which life forms may be preserved in materials of volcanic origin—as witness, especially, the fossils preserved in deposits formed from volcanic dust and ash in the vicinity of an eruption. The remains at Pompeii are among the most spectacular of this kind, but many other regions also have been explored for fossils buried under clouds of hot dust and ashes subsequent to some volcanic activity.

Metamorphic rocks, as the name implies, are those produced in the interior of the Earth, under conditions of high temperature and pressure, from igneous or sedimentary parent materials. The particular parent rock determines the kind of metamorphosed endproduct that forms, such as marble from limestone, slate from shale, and so on. Fossils would not be preserved, even if present, in the original parent materials, and any extant life would be unrecognizable if deposited in rocks formed under such extreme physical conditions.

Sedimentary rocks form by the accumulation of rock particles cemented together to produce a new kind of material, usually under water in relatively quiet circumstances. Various sorts of parent materials may be altered during the formation of sedimentary rocks: for example, sandstone was formed from granite particles cemented together primarily by minerals of the iron and magnesium groups, shale from petrified mud, and chalk from the calcareous remains of countless numbers of protozoa of the Foraminifera group. The white cliffs of Dover on the English coast provide a dramatic example of chalk deposits, and the sandstone cliffs of the Grand Canyon and other regions of the southwestern United States are breathtaking sights that are rich in fossils.

The Fossil Record

A fossil is *any* evidence of former life, including excreted metabolic deposits and impressions (such as footprints) which exist without the remains of the organisms that left such artifacts behind. General requirements for fossilization include a number of factors, but the most important for most life forms are (1) rapid burial, which leads to less chance of destruction by the actions of scavengers and predators or by oxygen-requiring decay processes, and (2) the presence of some hard parts more resistant to destruction, including such structures as shells, bones, and similar animal components, or the woody water-conducting vascular tissues of many land plants. The combination of particular requirements for successful fossilization is such that the great majority of organisms never take the first steps leading to preservation. The probabilities are very low indeed that all the factors will mesh together in the right way at the right time in the right place.

The usual prelude to fossilization involves burial in sediments of aquatic zones, such as oceans, lakes, seas, ponds, streams, and so forth. Bodies settle to the bottom of such watery graves and become covered by sediments or, on rare occasions, by volcanic ash and dust spewed over and into the aquatic zone. There is little scavenging or predation, and decay occurs slowly and incompletely under the usual conditions. The hard parts persist or become mineralized after the soft parts are removed or carried away. Sediments continue to accumulate, and lower strata solidify while additional layers pile up for as long as suitable conditions persist for such events. Sedimentary rocks of marine environments typically are stratified in this way and can be recognized long afterward when water no longer is present in that region. The steep sides of the Grand Canyon show just such a sequence of stratification of a previous aquatic zone.

This sequence of events permits us to understand and even to predict the existence of gaps and distortions in the fossil record. We would expect that many more aquatic than terrestrial forms would be preserved and that organisms lacking hard parts would be rare components of the fossil record. In view of the fact that the fossil record represents a sampling of former life, we also would predict that rock formations would have a richer representation of fossils from larger populations than from smaller ones simply because of the low probability of a rare event's actually taking place. It is more likely that some members from numerous or large populations of a species would have become fossilized than that some individual from a scarce or a small population would have been preserved in the fossil record. We would expect a richer fossil deposit when the climate is warmer since more of the land would be covered by shallow seas suitable for the preservation of life forms. During cooler, drier times more land area is exposed to erosion and less water is available for sedimentary rock formations in which organisms could be fossilized. The relative changes in the levels of land and sea areas, as a function of climatic changes, produce variable conditions for fossilization. When more land area is exposed there would be more kinds of habitats and thus more varied selection pressures acting to fix variation in the gene pool; therefore there would be more kinds of life forms in such habitats. But populations probably would be smaller in size at such times of flux, and presumably they would be more transient during changing environmental phases. For these reasons we would expect even greater sampling distortions in the fossil record of the diverse life forms that existed during the more dynamic intervals in geological history. Generally then, the fewest *kinds* of forms are likely to become fossilized when conditions are best for preservation, because fewer kinds of species would exist, though in relatively large populations; and when conditions are poorest for fossil formation there would be many kinds of species but fewer samples preserved because of smaller populations and the unfavorable conditions for sedimentary rock formation and for fossil deposition, especially for terrestrial life. Discontinuities and distortions therefore are predictable characteristics for the fossil record of ancient life.

From the foregoing considerations we would predict that some groups would be fairly well represented, while others would be inadequately preserved in the fossil record. Invertebrates and protists that secrete calcareous or siliceous hard parts would be more likely to fossilize—for example, the corals of the coelenterate group and the Foraminifera among the protozoans. Among the invertebrates we find a good record for some of the wormlike and annelid phyla whereas others virtually are missing; there are many marine arthropods but very few fossil insects; and there are numerous speciments of brachiopods (lamp shells), molluscs (clams, snails), and echinoderms (starfish and allied forms). Among the vertebrate animals there are substantial numbers of most groups but very few protochordates (such as *Amphioxus*). There are very few fossil birds, but many amphibian and reptile remains; and fewer primates than other kinds of mammals are represented in the fossil record. For plants we find very few nonvascular forms such as algae, fungi, and bryophytes (mosses and liverworts), but a fair amount of the woody species of vascular plants. Unfortunately the softer portions of woody plants rarely are preserved; thus identification often proves to be difficult or impossible in cases where flower parts are required for determining relationships and classification. Very often the fossil classification is based on remains of plant spores and pollen, and these provide a relatively inadequate basis for interpreting relationships or evolutionary trends. In some cases stem and root parts of the same plant have been given different formal names at first, a revised and more accurate identity becoming possible only when such parts happened to be found attached as a single individual, at some later time.

SPECIAL INSTANCES OF FOSSILIZATION

We already have mentioned one of the unusual ways in which fossils may be preserved, namely, burial under hot volcanic ash and dust. There is rapid dessication under these conditions as well as preservation in unusually complete form, and destructive factors ordinarily do not intrude. Beautifully preserved specimens of such fragile species as insects may be found in areas exposed to such volcanic activity. Organisms may have fallen into the water below as they became covered with volcanic ash and then become fossilized under the ordinary conditions for sedimentary rock formation. The ruins and the fossils at Pompeii provide spectacular evidence of the considerable detail that may be retained in materials that are rapidly buried under volcanic debris itself. Many excellent specimens of insects have been found encased in amber, which forms upon hardening of the sticky resin exudates from various trees including pines and other conifers. The detail is so well preserved in some cases that it is possible to identify the insect species and find that many millions of years may have passed without substantial change when compared with the same insect in modern times. The petrified wood found in the southwestern United States and other regions represents fossil plant remains in which minerals were deposited alongside organic tissue components in such a way that one can obtain "peels" of exposed surfaces and see the cellular architecture in preparations viewed under the microscope. A dramatic example of special conditions leading to fossilization is that of burial in petroleum springs that developed into sticky tar pools. The volatile oils evaporated from the petroleum to produce tar, which later changed to viscous asphalt. Animals and other life forms that became trapped in the sticky tar are very well preserved and can be recovered intact in many cases. The La Brea tar pits in Los Angeles, California, are famous for the wealth of birds and mammals from the Pleistocene and Recent epochs of Cenozoic times. Another bizarre fossil-forming situation involves the quick freezing of tundra-dwelling animals, which are so extremely well preserved that tissue specimens can be examined easily and in great detail.

Many specimens of reptiles and mammals have been excavated from the frozen tundras of Siberia, and some wild tales have been told about such fossil animals. It was claimed by some Soviets that frozen lizard fossils were thawed out after thousands of years in the ice and found still to be alive. These reports were never substantiated by independent tests performed by others, and so they have been taken with a grain of salt by the scientific community. Frozen fossil lizard appeared on the luncheon menu for an Explorers Club meeting a few years ago, and the gourmet verdict was that it tasted terrible.

At various times it has been assumed that catastrophes must have occurred to cause devastation among the huge herds of Pleistocene mammoths, thus accounting for the great numbers of these large hairy creatures that ended their existence as well-preserved frozen fossils. Yet one careful analysis revealed that only 38 specimens of frozen mammoth had been described by 1960, whereas calculations from the total fossil record indicated that a population of about 50 thousand must have existed during the times when the frozen animals were buried. This discrepancy alone was sufficient to cast doubt on the idea of catastrophe as the source of the frozen mammoth fossils. Further evidence against the occurrence of catastrophe came from a study of the 38 specimens. It was revealed that a quick burial was unlikely for 34 of the 38 creatures, because there was substantial evidence of damage from predators or scavengers. The remaining 4 mammoths were very well preserved and essentially undamaged, but all of them showed clinical symptoms indicating that they had died of suffocation. Examination of their stomach contents indicated that each had eaten well before death and that each had been in good health. The fact that some slight decay was evident pointed to a burial that probably was not immediate for any of the 4 fossil specimens.

In addition to mammoths, various other species of large animals have been found frozen, including the wooly rhinoceros. In fact, the species that have been found as frozen fossils are not a random sampling of the life forms that inhabited the region at that time. There is instead a preponderance of rather large and somewhat clumsy animals. In view of this fact, together with other observations that have been made, the most reasonable interpretation of the frozen fossils is that they represent those occasional unfortunates that were trapped somehow after straying too close to a dangerous situation. Although this is not catastrophic at the species level, it certainly was catastrophic for the individual animal caught in such a death-trap. The fossils from the La Brea tar pits provide a similar image of a collection of accidental strays that blundered into the pools and became trapped there, along with such predators as the saber-toothed tiger, which may have gone after the trapped prey and been rewarded for their greed by the same deadly fate that befell their intended victims.

CAN WE USE SUCH A DISTORTED
AND FRAGMENTARY FOSSIL RECORD?

Despite its inadequacies, the fossil record fits the predictions that can be made on the basis of the varied comparative studies of living species groups. There is a continuum of relationships among fossil and living forms, which is explained most simply by genetic and evolutionary lineages (Fig. 10.3). The time sequence of the fossil record is such that the most ancient deposits contain only the simplest life forms, and progressively more complex life appears in progressively younger strata. Prokaryotes are the first organisms to have been deposited; eukaryotes did not appear until much later, and the simplest eukaryotes occur in the oldest rocks. Complex plant and animal forms do not appear until a still later time. Among the complex forms, there is a predictable sequence of vertebrate life in the fossil record, with the earlier deposits encasing primitive fishlike vertebrates and the youngest formations containing mammal species. The time sequence is consistent. The fossil flora and fauna of a particular period generally resemble others of the same or nearby periods of time, and many of the most recent fossils are similar or identical to modern life forms. Such consistency implies descent with modification, not the operations of random processes of species formation over the long periods of the Earth's history.

Careful study of fossil forms themselves and of present-day life reveals similarities that have a genetic foundation, although some resemblances are clearly superficial and unrelated. It is important to distinguish between *homologous* and *analogous* habits and structures if we are to understand phylogenetic relationships. We conceive of evolutionary processes generally as involving initial *divergence* of descendant lineages from common ancestral stocks. Subsequent to such divergence, later evolutionary events may

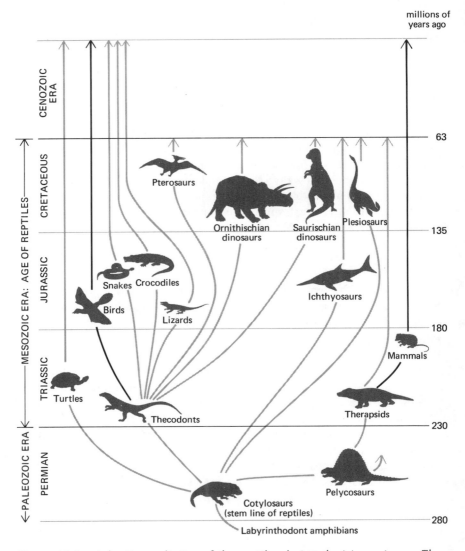

Figure 10.3   Adaptive radiation of the reptiles during the Mesozoic era. The position of each group in the diagram corresponds approximately with the time of greatest abundance for extinct forms or the time of origin of living groups. (Adapted from *The Dinosaur Book* by E. H. Colbert, McGraw-Hill, 1951, p. 52. Copyright © McGraw-Hill Book Company. Reprinted by permission of E. H. Colbert and the McGraw-Hill Book Company.)

produce *convergence* or may lead to continued accumulations of increasingly different life forms (Fig. 10.4). During divergence a variety of different life styles may develop among highly successful groups of species, each leading toward adaptation to a set of living conditions as a function of natural selection acting on diverse gene pools. Some one or more of these divergent lineages may come to resemble an earlier ancestral form, in which case we have evidence of the analogous development of traits during an episode of convergent evolution. For example, the reptilian ancestor was a terrestrial animal. Among the many kinds of divergent descendant groups that appeared during the Mesozoic era, there were aquatic and aerial reptile forms. The aquatic icthyosaurs superficially resembled the earlier aquatic fish ancestries, but these swimming habits and the successful anatomical and physiological traits which are associated with an aquatic existence were developed in entirely different ways in fishes and in icthyosaurs. The aquatic habit in these forms and in the aquatic mammals, such as whales and dolphins, provide examples of convergent evolutionary development. The structures are analogous and bear little or no genetic relationship to each other. The flying habit of insects, pterosaur reptiles, birds, and the

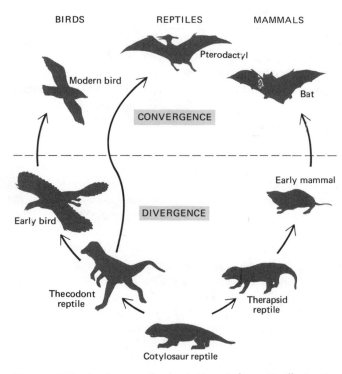

BIRDS   REPTILES   MAMMALS

Pterodactyl

Modern bird

Bat

CONVERGENCE

Early mammal

DIVERGENCE

Early bird

Thecodont reptile

Therapsid reptile

Cotylosaur reptile

Figure 10.4   Pathways of adaptation. Schematic illustration of relationships between the evolutionary phenomena of divergence and convergence, for selected adaptive modes of life.

mammalian bats provides a similar instance of convergent evolution of analogous structures. The formation of the wing, the organ of flight, probably occurs by completely different developmental pathways in these forms and thus has a different genetic basis.

Striking similarities have been described for the marsupial mammals that evolved in Australia and placental mammals that evolved elsewhere in the world during the many millions of years when Australia was separated from other major land masses. The various marsupial forms that inhabited living zones similar to those inhabited elsewhere by similar placental mammals include the doglike Tasmanian wolf, the bearlike koala, the woodchucklike wombat, and other amazing types. It is uncertain whether all such forms represent convergences in evolution. It is entirely possible that some of the similar marsupial and placental species evolved along similar lines because of mutations in similar genes subjected to similar selection pressures. If this were the case, then some species pairs could be considered to be examples of *parallel* rather than *convergent* evolution. The basic differences are that species with similar genetic backgrounds may undergo similar selection pressures, which fix similar adaptations into the non-interbreeding groups, as in parallel evolution; or that similar selection pressures acting on different genetic backgrounds may fix adaptations of superficially comparable traits in divergent populations. Genetic analysis is very difficult to perform with species that do not interbreed, and in the absence of compelling evidence the tendency has been to account for resemblances between distantly related groups as convergences rather than as parallels. This choice is not entirely arbitrary; the comparisons that have been made between living forms usually have shown that the apparently similar anatomical, biochemical, physiological, or other genetically-based features are indeed the outcome of convergence during evolution in the vast majority of cases.

There is abundant evidence of homologies. In these cases we observe what appear to be very different structures, behaviors, and life styles, but comparative analyses clearly show the genetic relationships that underlie the basic plan common to the superficial differences, and the pathways of derivation. Thus among the mammals it would seem that the arm of a human, the foreleg of a horse, the wing of a bat, and the flipper of a whale should be completely different sorts of structures, since each represents an adaptation to a very different and distinctive locomotor function. Yet, though clearly very different morphologically, they are all remarkably similar anatomically (Fig. 10.5). The skeletal plan of the structure is essentially the same in every case, but in each case modifications have occurred that are particularly well suited to a different function. The scale of the reptile has become modified as the feather in the birds, which trace their ancestry to a particular reptilian stock. The two structures, scales and feathers, are adaptations to quite different functions but have a common genetic program for development during ontogeny.

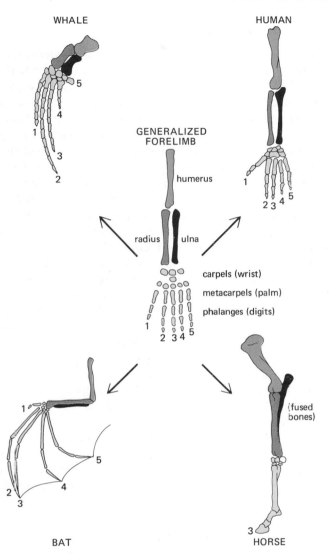

WHALE

HUMAN

GENERALIZED
FORELIMB

humerus

radius   ulna

carpels (wrist)

metacarpels (palm)

phalanges (digits)

(fused
bones)

BAT

HORSE

Figure 10.5   Homologous forelimbs of selected mammals. Each limb is
adapted to a different functional role in the species.

Homologies are examples of divergent evolution, which may seem anach-
ronistic, and analogies are the products of convergent evolution. In every
case that can be analyzed appropriately we can interpret the underlying
similarities of genetically controlled traits in terms of evolutionary descent
with modification from ancestral forms to the divergent or convergent
lineages derived from them. The same patterns can be discerned among
living groups as among fossil forms or for fossil and living species in
phylogenetic sequences.

IS EVOLUTION A "STRAIGHT-LINE" PROGRESSION?

Some fossil lineages show an apparently undeviating progression of continued improvements from an initial ancestor to its modern descendants. Such information led to the proposal that species evolve toward some particular end point of perfection. This principle of orthogenesis implies a strong component of metaphysics or vitalism or purpose in evolution. There is no substantial evidence to support orthogenesis or its philosophically oriented premises. The fossil record provides instead rather clear evidence that evolution has proceeded along reticulate pathways such that divergent lineages and evolutionary "dead ends" occurred alongside some one or more progressions that produced increasingly successful groups of species.

An especially well-documented case history of reticulate evolution has been described for the horses of the Cenozoic era, which evolved from the five-toed *Eohippus* to the present-day one-toed *Equus* (Fig. 10.6). There were numerous sidelines, which became extinct, as well as a major lineage in which changes led from the terrier-sized, five-toed, browsing *Eohippus* to the larger, one-toed, grazing species of *Equus*. The modifications represent adaptations to new habits and living conditions, adaptations that undoubtedly arose as a consequence of random mutations fixed by selection pressures, and not from some inner mystical force driving the species toward logical improvements. If there were such an orthogenetic principle, then it would be difficult to explain the divergent species and the lack of success of some descendant lines in this same group of animals.

Evolution is unpredictable and also opportunistic. Some of the major improvements in life forms may have had relatively little significance in the ancestral forms, or may have fulfilled quite different functions, yet they came to specify very important traits in a new species group. The mammary glands are modified sweat glands, each type serving quite different secretory functions, which are altered from their original nature in the early mammals and presumably in the immediate reptilian ancestor. Similarly the distribution of hair or fur in mammalian skin bears a relation to the presence of sebaceous glands as well as hair follicles. The insulating covering of fur, which is a distinctively mammalian trait, may represent also an opportunistic development based on modifications in the kinds and distribution of glands in mammalian skin, compared with reptilian traits. The skeletal gill supports of jawless vertebrates became modified, the first of the six supports developing into the jaws of the first fish descendants of the ancestral agnathans and subsuming a very different function. Many examples could be cited, all showing the opportunism that underlies many of the most significant improvements in the evolution of more successful life.

THE IRREVERSIBILITY OF EVOLUTION

A particular life form has never reappeared once it has become extinct. This observation is understandable in general genetic terms. Each new line in evolution has developed from a preexisting ancestral form as a

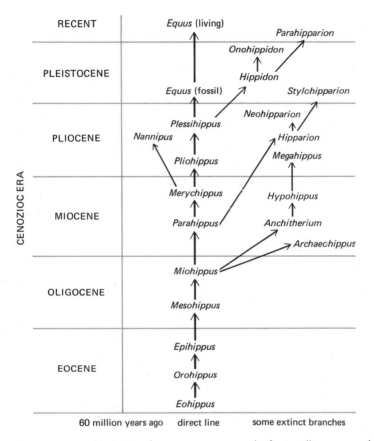

Figure I0.6   Horse evolution over a period of 60 million years during the Cenozoic era. In addition to the ten major forms in the direct line from *Eohippus* (also known as *Hyracotherium*) to *Equus*, various evolutionary side branches occurred and became extinct at different times.

consequence of numerous small mutational steps, each of which was incorporated into an extensive genetic program. Not only was each new mutation subject to a particular set of selection pressures, but each mutation in the sequence became fixed toward adaptation as a result of numerous selection pressures over long periods of geological time. Since such alterations are going on all the time, it would require a precise set of reverse mutations or new mutations similar to the original expressions for the prior sequence of steps to be retraced. Because of the improbabilities of mutation and selection pressures identical to those in past geological times, and in correct sequences, the reappearance of an extinct form is extremely unlikely and probably impossible. The improbability of identical environments at different geological times further compounds this unlikelihood.

Adaptive
Radiation

The rapid and extensive divergence of a single ancestral line into a number of descendant lines, which invade different habitats and exploit varied modes of life during their continued evolution, has been well documented in the fossil record. Such adaptive radiation may involve limited groups of species, such as the Darwin's finches, which we will discuss shortly; but more often it refers to events concerning whole classes and subclasses of organisms. This phenomenon permits the study of long-term trends and of ecological divergence during evolution. The primary adaptive radiation often is followed by one or more secondary radiations, which exploit more fully some set of life styles produced during the initial divergence episodes.

Among the placental mammals the primary radiations emanated from the small terrestrial insectivores during the late Mesozoic and early Cenozoic eras (Fig. 10.7). From these ancestral insectivore types radiation led to divergent lines adapted to varied living zones that had not been exploited by the ancestral stocks. *Aquatic forms* developed from three different terrestrial lines: (1) a carnivore group, which produced the flesh-eating whales and porpoises, (2) another carnivore ancestor that diverged to the seals and sea lions, which are aquatic at all times except during reproduction, and (3) a herbivore line, which led to the manatees and sea cows. *Ground-living forms* diverged to five principal groups: carnivores, rodents, elephants, odd-toed ungulates, and even-toed ungulates. *Arboreal forms* are found principally among the primates. And *aerial forms* include the many kinds of bats.

Secondary radiations involve continued diversification leading to saturation, in some cases, of a particular life style or adaptive zone. The bats have thus diversified to include fruit-eaters, insect-eaters, fish-eaters, bloodsuckers, and other types. Whales now include the giant whalebone food-strainers; carnivorous whales, which prey on other marine mammals; and big-toothed whales, which are adapted to feeding on deep-sea cuttlefish. The ground-living edentates of South America diverged to heavily armored armadillos, the ant- and termite-eating anteaters, arboreal sloths, and sluggish herbivorous ground sloths, which now are extinct.

Other examples of adaptive radiations include the divergence of the reptiles of the Mesozoic era to include aerial forms (pterosaurs), aquatic forms (ichthyosaurs), herbivores, carnivores, erect running bird-like forms, quadrupedal running types, arboreal forms, the many kinds of huge dinosaurs, and other types, which have left a rich fossil record (see Fig. 10.3). The radiations of Australian marsupial mammals produced numerous arboreal species, very few carnivores, and no truly aerial or aquatic forms.

There are some excellent examples of localized adaptive radiations

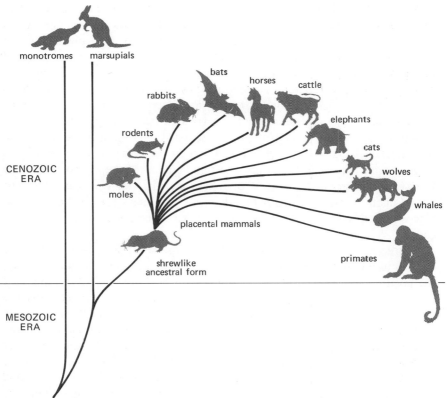

Figure 10.7   Adaptive radiation of the placental mammals during the Cenozoic era. (Adapted from *A Biology of Human Concern* by W. Etkin, R. M. Devlin, and T. G. Bouffard, J. B. Lippincott Company, 1972, Fig. 9-26, with permission of the publisher.)

among some groups of invertebrates, fish, and birds: the gammarid shrimp of Lake Baikal in Siberia, the cichlid fish of Lake Tanganyika in Africa, the sicklebill birds of the Hawaiian Islands, and Darwin's finches of the Galápagos Islands off the coast of Ecuador, among others. This group of finches occurs nowhere else in the world but on the volcanic Galápagos Islands in the Pacific Ocean, about 600 miles off the western coast of South America. These islands are about 1 or 2 million years old and never were connected to the mainland or to the islands of Polynesia to their west. During this brief interval of geological time some 14 different species evolved from a presumptive ancestral finch that probably arrived from the South American mainland, where distant relatives still can be found. One of these 14 species is found exclusively on Cocos Island, which is 600 miles northeast of the Galápagos and is not a part of that volcanic chain. The Cocos Island finch is clearly related to the 13 species of the Galápagos Islands, and all together they comprise the subfamily Geospizinae of the fringilloids or finch family. The radiation was such that 6 species evolved into ground finches, which feed on seeds of different sizes and on the fruit of the prickly-pear cactus; 6 other species are tree finches, which have diversified into 3 kinds of insect-eaters and 3 unusual types of finch. One of these latter is a vegetarian with a parrotlike beak; another is an insect-eating species, which inhabits mangrove swamps; and the third is the unique woodpecker finch, which probes for insects in the holes of trees by using a twig or a cactus spine as a tool. The remaining 2 of the 14 species are a warblerlike finch on the main island chain and the single species of Cocos Island finch. It is an interesting fact that only 1 species developed on isolated Cocos Island whereas 13 species diverged on the islands of the Galápagos chain.

There are many unusual aspects of the fauna and flora of the Galápagos Islands in addition to Darwin's finches, and Darwin himself commented on the peculiar life forms and their variations in his book describing this port of call and others during the long voyage of the H.M.S. Beagle. Besides there being unique forms of life on these islands—such as the marine iguana, giant tortoise, and cactus trees—the islands are unusual in lacking native forms of amphibians, palms, plants of the lily family, and a number of tropical plant families that ordinarily would be expected on land areas situated at the equator. Island populations and unique displays of island species provided both Darwin and Wallace with a great deal of material in support of their theory of evolution by natural selection, and against the idea of special creation.

Mass Extinction

In contrast with the disappearance of some particular species or group of species during geological history, mass extinctions involve the relatively sudden disappearance of large groups of dominant forms after many millions of years of successful and widespread existence. The previously popular notion was that catastrophes were responsible for such extinction episodes, especially in view of the suddenness of the disappearances from the fossil records. Because of the gaps and discontinuities in the fossil record, which often coincide with mass extinctions, we know little about the duration of some of these episodes and less about the mechanisms responsible for these mysteries.

Of the 2500 families of animals known from the fossil record, only about one-third are still living. Most of the others simply dropped out of sight and left no descendants, although a few did evolve toward new life forms. There is little evidence for mass extinctions of plants, but this is not altogether surprising. There is little correlation between major changes in fauna and flora, according to the fossil record. Each of the three principal land flora (mosses and ferns, gymnosperms, and angiosperms) apparently arose during brief incidents of rapid evolution and then remained stable for many millions of years. Among the fossil records for animals, on the other hand, there are some episodes of mass extinction, which provide reference points to denote boundaries of geological time. About two-thirds of the 60 different families of trilobites became extinct at the end of the Cambrian period. These marine arthropods, distantly related to some modern crustaceans, were worldwide in distribution and thus provide an abundant and useful index for Cambrian strata.

The mass extinctions affected many unrelated groups of organisms from a variety of living zones and occurred principally at the ends of the Cambrian, Ordovician, Devonian, Permian, Triassic, and Cretaceous periods (Fig. 10.8). At or near the close of the Permian period of the Paleozoic era, nearly half the known families of animals throughout the world became

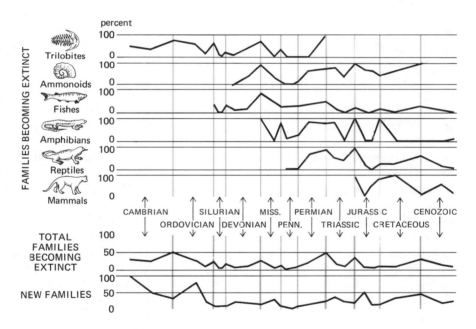

Figure 10.8   Patterns of extinction of some animal groups. Times of more extensive episodes of extinction are indicated by heavier vertical lines. (Adapted from "Crises in the History of Life" by N. D. Newell, *Scientific American*, February, 1963, p. 84. Copyright © 1963 by Scientific American, Inc. All rights reserved. Reprinted with permission of the publisher.)

extinct. This was one of the "worst" episodes of mass extinction; a comparable level of animal diversity was not restored until about 15 to 20 million years later in the Triassic period. Various major groups dropped out throughout the Permian period, but many others survived long enough to go out together at the end of this period. Although 75 percent of the amphibians and more than 80 percent of the reptile families disappeared during the Permian, the main suborders survived the Paleozoic era and continued into the Mesozoic, as can be seen from the Triassic fossil deposits. The previously predominant primitive amphibians and reptiles were replaced by the early dinosaurs, which were widespread by the close of the Triassic period. By the end of the Cretaceous period, which ended the Mesozoic era, about one-fourth of the known animal families had become extinct although the plants of that time were essentially unaffected. Among the animals of worldwide distribution that disappeared at the end of the Mesozoic were the dinosaurs, marine and flying reptiles, and the abundant ammonite and mollusc groups. Many plant and animal families persisted into the Cenozoic era, including fishes, mammals, and the turtle and crocodile groups of reptiles (see Figs. 10.1 and 10.3).

A number of extinctions marked the end of the Pliocene epoch during the Cenozoic era, and some of the more recent and dramatic episodes occurred in North America during the Pleistocene epoch, which followed. About 11,000 years ago, when the continental glaciers were at their greatest extent over North America, a very rich and varied fauna of large mammals occupied the ice-free regions of the continent. There were many species of bears, horses, elks, beavers, and elephants, which were much larger than their modern relatives. This diversity of animal life was comparable to the striking variety of animal species in the sub-Saharan region of Africa today. As recently as 8,000 years ago there were horses, elephants, members of the camel family, and many other placental mammals on all the continents of the world except Australia and Antarctica. Most of these groups today are confined to small regions on one or two continents. Most of the species disappeared from North America between 6,000 and 12,000 years ago, with a peak rate at about 8,000 years ago. Most of the extinctions took place when the climate was fairly mild and the glaciers were retreating, rather than during the hard times of glacial advance. Furthermore, there were no such mass extinctions in Africa at the time that such events were taking place in North America and, more gradually, in Asia and Australia. Many of the large mammals of that time, both herbivores and carnivores, became extinct in just a few hundred years even though they had previously inhabited regions of varying climates around the world. Few species other than these large mammals seem to have been affected in any significant way during this episode of extinction. Radiocarbon datings have indicated that these now-extinct large mammals of North America began to disappear first from Alaska and Mexico and later in the Great Plains region of the United States. It is entirely possible that the last survivors still were present in Florida as recently as 2,000 to 4,000 years ago. The ecological habitats vacated by these extinct species, which included about three-fourths of the North American herbivores, remain unfilled today.

The extinctions during the Pleistocene are particularly baffling just because we have so much information about climate and other factors. Glaciation per se probably was not a primary factor; most of the extinctions occurred during times of final melting and glacial retreat, *after* the many species had successfully withstood several glacial and interglacial cycles. Furthermore, the glaciers did not cover all of North America, so that the southern latitudes were less directly influenced by the ice sheets than the northern latitudes, where they were present. Climatic changes took place during these extinction incidents, with higher mean temperatures and drier conditions during times of glacier retreat. Some forests were replaced by deserts and steppes, but no major habitat was obliterated during the Pleistocene. Extinction of species in the tropics certainly cannot have been affected directly by the glaciers in the north, since tropical climates at that time were essentially the same as today.

Was man, the fearsome predator, responsible for any of these extinctions? Agriculture became established in some human cultures perhaps 7,000 to 10,000 years ago, replacing a total dependence on hunting and gathering of food in these societies. Many cultures continued to be principally hunters and gatherers, even to the present day. Humans certainly were present on all the continents, except Antarctica, during the Pleistocene mass extinctions, but they had been present in these same areas for some thousands of years prior to this. Human societies were in fact more recent and scattered in North America than in Africa or Eurasia during these same times, but mass extinctions did not occur elsewhere to the same extent as in North America between 6,000 and 12,000 years ago. Still, these episodes of mass extinction did take place in waves that coincided with the extended range of expanding human societies and the migration of peoples to more comfortable or newer climatic regions. The suggestion has been made several times that many of the extinctions were a consequence of "overkill" by human hunters. This certainly is the case today, but the modern motives often are economic, leading to wanton destruction of life; recent extinctions are not due to orgies of killing for sport or for thrills. It is "civilized" human beings, not the indigenous inhabitants of a region, who have been responsible for the extinction of hundreds of animal species in the past few hundred years. The observed correlations of mass extinctions and the migrations of early human societies with a change from cool, moist to warmer, drier climates should not be ignored. But whether there is a cause and effect relationship still remains to be determined.

Extinctions of single species generally occur by quite different pathways, except for the deliberate murder of some species and the squeezing out of others by human activities recently. The usual situation involves two particular sets of events. The size of breeding populations may become reduced because of genetically low adaptive value of species features, until new progeny no longer are produced to perpetuate the species. Such populations may be more susceptible to predators and so enter into a decline, or they may fail to secure all their needs for continued existence, for any one or more reasons. Whatever the reasons, they lead to lowered adaptiveness of the gene pool in a particular selective environment. Their adaptability is limited, perhaps because of too high a level of adaptedness (see pp. 144–145). Other species may become extinct by virtue of evolving into new species subsequent to genetic modifications under the influence of natural selection. The more likely situation is one in which some population of a species evolves toward a more successful genetic type, which continues to change until a new species level is achieved, while the remaining populations of the original species decline to the point of extinction because of reduced probability for success in the presence of the new and more highly adapted population. Conventional processes are more than adequate to provide the theoretical basis for an explanation of species extinctions, unlike the enigma of mass extinction as a phenomenon of biological evolution.

Continental Drift

Until recently, the generally accepted view of the Earth's topographical history was that of a crust that remained essentially unchanged after its initial cooling more than 4.5 billion years ago. The changes that can be observed in the geological record were assumed to have occurred because of climatic, geological, and biological factors. The concept of migration of the continental land masses had been proposed at various times during the past few centuries by a number of authors, including Sir Francis Bacon, but the idea had never been accepted. The obvious resemblance of the existing continents to neatly fitting pieces of a jigsaw puzzle (Fig. 10.9) permitted even the most casual observer to comment on the coincidences. During the latter part of the nineteenth century, Eduard Suess fitted together the lands of the southern hemisphere and suggested that the conglomerate region be called Gondwanaland (named for Gondwana, a key geological province of east central India). Antonio Snider had suggested earlier that all the continents originally comprised a single land mass. His suggestion came from attempts to explain the amazing similarities among fossil materials from widely separated areas of the present world.

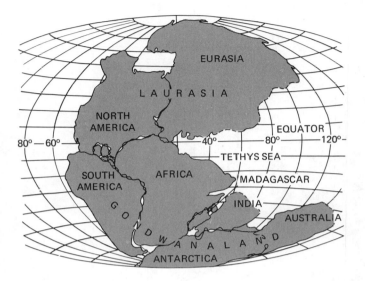

Figure 10.9   Pangaea. The supercontinental land mass of Pangaea as it may have been distributed during the Permian period of the Paleozoic era.

The principal theory of continental drift was first proposed by Alfred Wegener in 1912 and expanded during the 1920s and 1930s to accommodate additional documentation he had amassed. Wegener proposed that there had been one giant supercontinent, which he called Pangaea, at a time prior to the Mesozoic era, more than 200 million years ago (Fig. 10.9). He suggested that this land mass began to rupture during the Triassic period to produce two continental masses. He named these Gondwanaland in the southern hemisphere and Laurasia in the northern half of the world, the two regions being separated by the Tethys Sea in which they were located. The theory and the data were dismissed and revived at various times until about 10 or 12 years ago when extensive data were presented in support of the phenomenon. Two central problems were examined in relation to the possibilities for migrations of huge land masses. First, if the crust and mantle of the Earth were rigid as supposed, then how could there be lateral movements of continents? The second problem concerned whether there was an energy source available for the presumed migrations.

Geological observations and accumulating lines of evidence have provided important insights into processes and events that are fully compatible with the theory of continental drift. The Earth's crust provides clear evidence of large-scale movements—periodic rejuvenation in the display of new mountain chains and ocean trenches; mixtures of older and younger

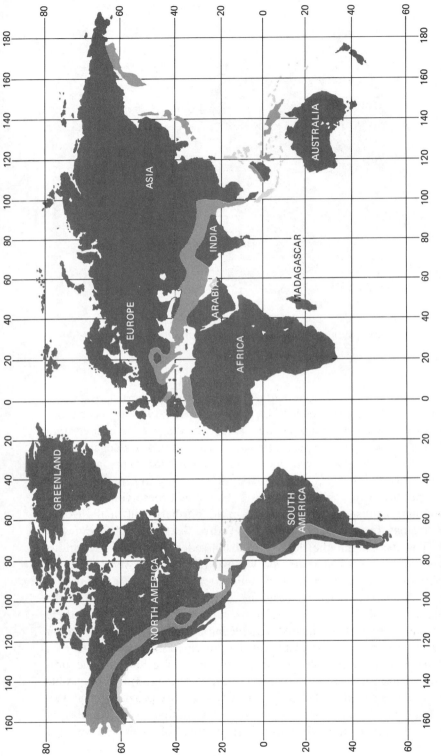

Figure 10.10 Global perspective of belts of geological activities. (Adapted from "The Confirmation of Continental Drift" by P. M. Hurley, *Scientific American*, April, 1968, pp. 56–57. Copyright © 1968 by Scientific American, Inc. All rights reserved. Reprinted with permission of the publisher.)

strata, which are indicative of warping and folding as a result of intrusions by younger formations into older deposits; and a globe encircled by belts of geological activities (Fig. 10.10)—all of which contribute toward the reality of the occurrence of vast motions of the planet's crust as reasonable explanation for the phenomena. Specific lines of evidence in support of the theory have come from studies of the topography of the ocean floor using sonic depth recordings, from analysis of magnetism patterns revealed in iron-bearing rocks of different ages and regions, from datings of geological formations presumed to have been in other locations in earlier times, from studies of the fossil record, and from continent reconstructions based on geological features of some of the significant land masses of the present day.

A sequence of events has been proposed, beginning about 225 million years ago at the close of the Permian period of the Paleozoic era. Pangaea began to split up, forming Gondwanaland to the south and Laurasia to the north (Fig. 10.11A). The fern forests, which fossilized to form our principal coal deposits, were widespread at that time, and reptiles already had evolved from amphibian ancestor stocks. While reptiles continued to evolve during the Triassic period, some 200 million years ago, Gondwanaland began to subdivide: the common land mass comprising South America–Africa began to separate from Antarctica–Australia (Fig. 10.11B). Separations continued through the Jurassic period until the final rupture of South America from Africa and of Australia from Antarctica, about 65 million years ago during the Cretaceous period (Fig. 10.11C). At this time mammals were just beginning to succeed the reptiles as the predominant animal life on the Earth. Since the Cretaceous period ended and the Cenozoic era began, about 60 million years ago, some of the land masses have continued to migrate considerable distances; others have remained more or less stationary. During the Cenozoic era, while Eurasia and Antarctica remained relatively stationary, except for rotation, India became a part of the Asian region of the Eurasian land mass, North and South America became joined with the uplift of the Central American land link, and the northeastern region of Africa joined the Eurasian mainland (see Fig. 10.10). Two of the major ocean trenches have remained relatively stationary. They originally lay west of the Americas but ended up east of these continents when North and South America reached their present positions: The continental migrations are still going on at this moment, and the face of the Earth will look quite different 50 million years from now, if we don't destroy it in the near future.

Some general conclusions have been made on the basis of evidence compiled during the past 10 or 15 years, in which continental drift and sea-floor spreading have been merged into a general theory of plate tectonics (Fig. 10.12). To a depth of about 100 kilometers (about 62 miles), the continental areas have greater strength than the ocean basins. The land areas therefore tend to maintain themselves as buoyant masses that can be

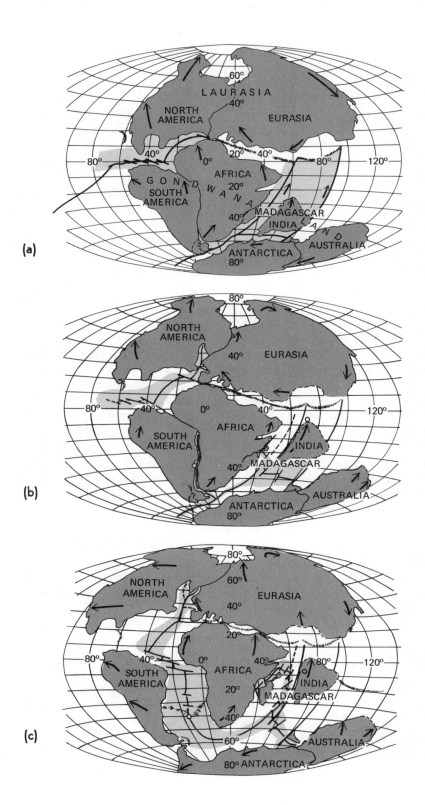

(a)

(b)

(c)

ruptured but not destroyed by sinking motions. Rising materials push surfaces apart while sinking materials pull surfaces together and toward the region of sinking. A sinking zone in the oceanic regions thus leads to continental movement toward that zone. If a rising zone is established under a continent, the land mass will split apart and the separate areas will move away from the rising zone, in different directions.

When the ocean floor moves toward a sinking zone in an oceanic region, it forms a deep trench bordered by volcanoes, chains of islands, or elongated land masses such as Japan or the Philippines. When an ocean floor moves toward a continent, it seems to pass under the continental border, forming a great mountain chain. The mountain chain may be composed partly of piled-up materials already present and partly of new and existing volcanic materials, plus the continental shelf and other areas. A mixed formation occurs basically by melt of the old and intrusion of the new materials. The western coast of South America appears to be such a mixture of old and new materials, as do the Himalaya Mountains, which presumably formed when India thrust itself into Eurasia. The contorted chain of mountain ranges from the western Atlas Mountains of northwestern Africa through the Mediterranean Sea, the western Alps, the Caucasus, and on to the Himalayas (see Fig. 10.10) is presumed to have resulted from the buckling of sediments in the Tethys Sea owing to continental movements at various times and collisions of land masses.

The outer shell of the Earth, or lithosphere, thus is viewed as consisting of a number of rigid tectonic plates, which are in constant relative motion, floating on molten or semimolten rock of the Earth's mantle. Such phenomena are believed to have occurred for at least the past 2 billion years, but some mechanism other than plate tectonics may have prevailed in still earlier times. The nature of the energy source is uncertain, but several promising lines of evidence have been collected. Although a considerable amount of information remains to be collected and deciphered, the concept of continental drift will not be discarded, as it was in the past just because all the answers are unavailable.

←——

Figure 10.11   Progressive fragmentation and redistribution of the world land masses during the Mesozoic era as a result of continental drift. Arrows indicate the probable directions of drift of the various portions of the land areas and their presumed locations (a) 200 million years ago during the Triassic period, (b) 135 million years ago during the Jurassic period, and (c) 65 million years ago during the Cretaceous period. Compare with Fig. 10.10 showing present-day distribution of the continents. (Adapted from "The Breakup of Pangaea" by R. S. Dietz and J. C. Holden, *Scientific American*, October, 1970, pp. 35–37. Copyright © 1970 by Scientific American, Inc. All rights reserved. Reprinted with permission of the publisher.)

The geological evidence has been supported very strongly by studies of the distribution of fossil species. For example, a particular fossil reptile species has been found in early Triassic deposits from Antarctica, southern Africa, and eastern Asia. The simplest interpretation of the occurrence of the same species in regions now thousands of miles apart is that these regions formed a common land mass during the Triassic, when this reptile species existed. Many other case studies of plant and animal fossils also support the continental drift proposals.

The consistency and logic of the framework provided by plate tectonics for various geological phenomena, and the successful testing of some of the theoretical tenets, have provided a solid foundation for the ideas that now are accepted by the great majority of the scientific community.

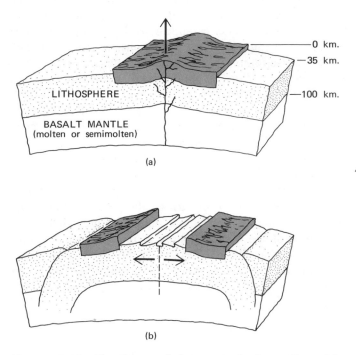

Figure 10.12   The theory of plate tectonics is an attempt to explain a mechanism by which continental drift may occur. (a) The process begins when a spreading rift develops under a continental land mass resting on a crustal plate that "floats" on molten or semimolten basaltous rock. Molten basalt spills out at the rift opening. (b) Drift also requires that the leading edge of the lithosphere plate descend into the mantle at a trench region. A new ocean basin develops between the two land masses. (Adapted from "The Breakup of Pangaea" by R. S. Dietz and J. C. Holden, *Scientific American*, October, 1970, p. 32. Copyright © 1970 by Scientific American, Inc. All rights reserved. Reprinted with permission of the publisher.)

Some Suggested Readings

Colbert, E. H., "Antarctic fossils and the reconstruction of Gondwanaland." *Natural History* 81 (January, 1972), 66.

Dewey, J. F., "Plate tectonics." *Scientific American* 226 (May, 1972), 56.

Dietz, R. S. "Geosynclines, mountains and continent-building." *Scientific American* 226 (March, 1972), 30.

Dietz, R. S., and J. C. Holden, "The breaking up of Pangaea." *Scientific American* 223 (October, 1970), 30.

Hurley, P. M., "The confirmation of continental drift." *Scientific American* 218 (April, 1968), 52.

Lack, D., "Darwin's finches." *Scientific American* 188 (April, 1953), 66.

Newell, N. D., "Crises in the history of life." *Scientific American* 208 (February, 1963), 76.

Weller, J. M., *The Course of Evolution.* New York, McGraw-Hill, 1969.

CHAPTER 11
SPECIATION

Casual observation of any living-zone on the planet reveals a diversity of life forms that can be distinguished from one another and that maintain their individualities generation after generation. The discontinuities between life forms are striking and baffling. Since gene exchange leads to a spectrum of variation in populations, how do differences arise that make one kind of organism so completely distinct from a neighboring species? Since these species arose from common ancestors of the distant past, how did the gene pools become so subdivided and differentiated with the passing of time? The two basic questions we pose in relation to species formation are (1) How did distinctly different groups arise from common ancestral stocks? and (2) once formed, how do they maintain their distinctness even when they inhabit the same living-zone? From what we have learned so far we would expect that differences arose by the processes of mutation and natural selection, which fixed adaptive variability in the gene pool of each species. If interbreeding leads to reassortment and recombination of genes with the resulting spectrum of graded variation, then we would further expect that interbreeding did not take place between present-day species. Furthermore, we would predict that interbreeding was prevented or reduced during species development, with consequent reduction or inhibition of gene flow between those evolving populations that were headed toward divergence.

Although it was entitled *The Origin of Species,* Darwin's treatise was more concerned with the inherent modifications in species and species groups and less with mechanisms leading to species divergence from a common pool of variability. The introduction of the concept of isolation as a crucial factor in speciation has had a profound effect on our understanding of the processes involved and has become an integral feature of neo-Darwinian principles of evolution (Fig. 11.1). In its simplest form, isolation may arise through any factor that separates an interbreeding population into units that are unable to exchange genes. The reduction or total cessation of gene flow between isolated subpopulations does not affect the continuing processes of mutation and selection. But different gene frequencies and genotype combinations may become established in these subpopulations because intergradation is prevented by the isolating factor. And as random mutations are acted upon by varying selection pressures, the most probable course of events over a period of time is that the genotype arrays in the subpopulations will become increasingly different. We would predict that, given sufficient time in isolation, divergence would continue to some stage at which the differences became great enough to identify different species derived from the common ancestral population. It is important to emphasize that isolation does *not* initiate divergence. It is simply that divergence is more likely to occur in separated populations because there would be extremely low probabilities for identical mutation and selection pressures to operate in different groups during the same period of time. For our purposes we will consider a species to be comprised of one or more populations that can interbreed and exchange genes regularly, or that have an undiminished potential for such interbreeding.

Geographical Isolation

It is generally accepted that the initial event in speciation is the physical separation of subpopulations by some external barrier that effectively prevents interbreeding despite the existence of such potential for gene exchange. It would be a simple situation to envision the imposition of some geographical barrier such as a new chain of mountains or an ocean separating parts of a population into disjunct units. Such massive features undoubtedly did impose barriers to interbreeding in some cases, perhaps during more dramatic moments of continental migration and the attendant formation of oceans and mountains. The separation of Pangaea into progressively fragmented land masses during the past 225 million years, especially during the Mesozoic–Cenozoic transition, has been invoked to explain the different evolutionary pathways taken by the mammalian fauna of Australia, South America, and other regions of the world in which the diverging groups finally became located. On a somewhat lower magnitude, the *Eucalyptus* species of Australia apparently originated in southwestern Australia during the Mesozoic era and spread eastward. These species became separated about 100 million years ago when eastern and western Australia were divided by a sea, and they remained isolated until about 1 million years ago when the land connection was reestablished. Divergence occurred during the intervening years, leading to differences sufficient to prevent interbreeding among the species even though they coexist in the same regions and habitats today.

There are other examples of speciation which, though the explanation is less obvious, can nevertheless be rationalized according to conventional evolutionary processes. There are about 300 species of gammarid shrimp in Lake Baikal and few species anywhere else in the world than Siberia. How

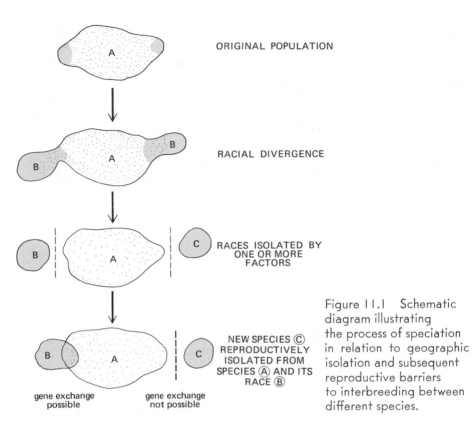

ORIGINAL POPULATION

RACIAL DIVERGENCE

RACES ISOLATED BY
ONE OR MORE
FACTORS

NEW SPECIES Ⓒ
REPRODUCTIVELY
ISOLATED FROM
SPECIES Ⓐ AND ITS
RACE Ⓑ

gene exchange
possible

gene exchange
not possible

Figure 11.1 Schematic
diagram illustrating
the process of speciation
in relation to geographic
isolation and subsequent
reproductive barriers
to interbreeding between
different species.

could speciation have occurred in a single lake, and how can these species remain distinct? Careful study has shown that there were numerous variations within the aquatic habitat, such as temperature differences, depth variation through one mile of water, differences in food sources and their availability, and different predators. It would fit the general theme very well if we could accept that divergence occurred initially when some subpopulations came to occupy various specified regions in the lake which others could not utilize. As genetic differences continued to accumulate, some populations might have extended their range, but divergence may have been achieved to a point that precluded interbreeding even though different populations came to be intermingled later on.

Among frogs of the species *Rana pipiens* in North America there is considerable variation among populations at geographically extreme ends of the total range, but there is a continuous graded series of populations that are intermediate in location and in the ability to exchange genes. The entire graded spectrum of phenotypes constitutes a *cline*, and in *Rana pipiens* there also occurs an interconnected series of populations which can exchange genes with their immediate neighbors but not with more distant populations from the same cline. One may view the *Rana* populations as a group of incipient species, since the obliteration of intervening populations would leave only those that were incapable of interbreeding to perpetuate the species as a whole. Continued differentiation would occur more rapidly because little or no gene exchange would occur and divergence would be more rapid when the populations were not connnected physically or genetically.

Isolating Mechanisms

During intervals of geographical isolation there would be genetic differentiation based on the fixation of adaptations of different kinds in gene pools of diverging populations. Some of these inherited variations would contribute to reduced gene flow between populations even if they inhabited adjacent living zones or the very same ecological situation. A great variety of genetic differentiations of life style has been reported, but we will illustrate the possibilities here with just a few examples.

There are many examples of closely related species that live in areas with somewhat different ecologies—or of races of a single species, such as the forest and prairie deermouse (*Peromyscus maniculatus*) groups. Among the deermice few hybrids are found in wild populations, but fertile progeny can be produced in the laboratory from matings between mice belonging to each habitat group. There is no genetic basis to prevent completely successful interbreeding once the opportunities are available. But there is a background of genetic differentiation of the two races, since each occupies a different living zone, which can be exploited successfully because of particular adaptations to each habitat. Ecological isolation is based on genetic modifications which, though they may not have led to strict barriers to interbreeding, as in the deermouse example, may in many other cases have been great enough to prevent effective gene flow even when hybridizations occur. Different species usually occupy different living zones when they are closely related, but often intermingle when the relationship is more distant.

Related species often do not interbreed in the wild, because they breed at different seasons or at different times in one season. Sometimes such species can exchange genes to produce fertile and viable hybrid progeny, as has been demonstrated under laboratory experimental conditions. One such example is provided by the toad species *Bufo americanus* and *B. fowleri,* which breed in the spring and often occupy the same ponds during the season. Few hybrids occur in nature, because the American toad breeds earlier and Fowler's toad breeds later, after *B. americanus* has completed its activities in perpetuating the species. Some hybrids may be produced in the wild if it happens that occasional individuals from each species are available and in reproductive condition. Differentiation of various traits clearly has taken place, since breeding seasons vary, appearances vary, and their calls are quite different (a trilling sound for *B. americanus* and a retching sound for *B. fowleri*), among other features. Since gene exchange is rare or absent in many locales, continued divergence would be expected to produce more profound distinction between these two species. Many related species of flowering plants also may be interbred in the laboratory though they fail to do so in the wild because of different times when pollen is shed and seeds are produced. As one watches the succession of flowering during a season it is quite usual to find that a species will come into flower after some relative has completed its reproduction, as occurs for goldenrods in the early fall, for woodland species of asters, and for many other examples.

In addition to ecological and seasonal mechanisms leading to isolation and consequent reduction in gene flow, a number of other modification patterns show the same general syndrome; that is, breeding between species is prevented by traits that make it unlikely that the different groups will be accessible to each other. Since the divergent species have little or no opportunity to exchange genes, each continues to evolve in its own particular direction of adaptation. Sufficient genetic differentiation ultimately could occur so that the populations would become reproductively isolated. At this stage they would be unable to exchange genes so that there would be no intermediate progenies and no blurring of species differences by backcrossing of hybrids with the parents.

Reproductive isolation may be evidenced in various ways, but three categories usually are distinguished. In each case there is no possibility of hybrid progenies contributing to species propagation; thus the gene pools of the parent species remain distinct and intact. In one set of cases hybrids are produced but fail to produce viable gametes in turn. Such sterile hybrids may be quite vigorous creatures, but because they are unable to reproduce, there is no further intermingling of gene pool components and no persistent effects from the hybridizations. A familiar example of such reproductive isolation is the hybrid mule, which is produced by matings between horse (mare) and jackass. The parent species are distinct, but there is a level of genetic differentiation of the chromosome complements so that the hybrid animals undergo aberrant meiosis and produce gametes that are nonfuctional. Failure of the chromosomes to pair at meiosis is symptomatic of genic differences between the paternal and maternal chromosome sets. Without proper pairing, chromosomes would be distributed erratically during the first anaphase of meiosis and lead to the formation of gametes with chromosome deficiencies and duplications. Such cells would be inviable, so the hybrid would be sterile. There are many examples of species hybrids that seem to have normal chromosome pairing at meiosis but nevertheless produce inviable gametes. In these cases it is presumed that genic differences reside throughout the chromosome complements and that the gametes are inviable because of genic distortions, even if the chromosome number in the gametes is correct. Studies of meiotic chromosome pairing and of hybrid sterility constitute a frequent approach to the analysis of genetic relationships between species.

The second evidence of reproductive isolation is that hybridization between genetically differentiated species may lead to progenies that are inviable; that is, the offspring do not live to achieve reproduction and therefore are incapable of perpetuating intermediate gene combinations. And, third, in crosses between species showing even greater differentiation, hybrids will not be produced at all. Such matings may lead to defective embryos or to the failure of fertilization itself or to any gradation indicative of the inability to exchange genes. In the vast majority of cases no attempts at hybridization take place. The distinctiveness of species that cohabit an area clearly is a function of matings that are intraspecific rather than of attempts at interspecific breeding which lead to failure.

Major Changes During Species Development

If we accept the foregoing discussions, then we can explain the origin of new species on the basis of the initiation of divergence by mutation and selection, and the amplification of such differentiation during isolation until the populations are reproductively incapable of interbreeding. Small changes, which accumulate, must be responsible for most of the observed variability in life forms. But there are magnitudes of evolutionary change that appear difficult to explain on this fundamental premise of cumulative small mutational steps leading to new forms. There are examples of major new traits and new groups of organisms that have appeared suddenly in the fossil record. The failure to find the "connecting links" or any substantial evidence of intermediate types showing gradual development of the new features of life forms has led to suggestions of other mechanisms or phenomena, which purport to provide the explanations for such novelties in evolution.

PREADAPTATION

The term *preadaptation* is used to refer to the phenomenon of chance adaptive variation whereby some mutations and genotypes may become advantageous in a new environment even though no observable advantage may have prevailed in the original selective situation. Such alleles or gene combinations may be retained if there is no great disadvantage when they are present in the gene pool, but they would not be widespread in populations unless selection pressures were altered. The suggestion is made that some populations may be preadapted, by chance, to a subsequently modified selection pressure and then become predominant under the altered living conditions. Rapid spread of a new trait, therefore, would be a consequence of the ready availability of suitable genetic materials in some populations and its high selective value under new conditions for the gene pool. Since these preadaptive traits would not necessarily be present in every population of a species, it might be that one or a few of the populations would undergo rapid change while others might be eliminated, might be sharply reduced in size, or might not change very much if selection was not particularly severe.

There are various examples of apparent preadaptive modifications in ancestral species which are detectably related to derived forms in which a new feature has been incorporated. Some of the best-documented studies have come from comparisons between blind cave-dwelling insects and their older relative species. The more ancient species exhibit characteristics that can be interpreted as preadaptive changes that have become stabilized and more distinctive in the descendant forms now inhabiting dark caves. The ancestral forms would have been preadaptively disposed to successful life in the dark if there had been chance variations affecting the vision apparatus; and if such types did encounter advantages in a new living zone, then they would have undergone rapid modification until they reached the particular endpoint we can detect today. In another example, there are strains of wild relatives of the cultivated potato species that are genetically resistant to some fungal infections, as can be shown by conventional breeding analysis. The assumption is that these are chance variants within the wild populations, which have no particular advantage in the absence of the disease-causing organisms. But should such fungi invade the region, then the populations which harbor the resistance alleles would be preadaptively disposed toward success while other populations of the same species may succumb to the infections. The many strains of antibiotic-resistant bacteria attest to this same principle of chance preadaptive variation, which leads to immediate success in an altered environment for those groups that happen to possess the alleles, and therefore such variants ultimately become predominant in the new environment.

MACROMUTATION

This category of inherited changes leads to drastic modifications, which may be of such magnitude that they immediately distinguish the mutants from the remainder of the population. These macromutations usually are mutations in a single gene, just like the commoner micromutations, but the consequences of the altered information may be quite substantial. Various mutants of *Drosophila* demonstrate major changes in anatomy or development. The members of the order Diptera, which includes all the two-winged insects, can be recognized by various traits, but the number of wings is invariable. The *tetraptera* mutant of *Drosophila* has four wings instead of two. The extra pair of wings form from structures that develop into halteres in normal flies. This mutant could be classified as a member of an entirely different order; therefore it would exemplify a kind of major mutational change leading to novelty in evolution except that the *tetraptera* mutation is harmful and the gene change in this case carries no evolutionary significance. Many families of flowering plants are classified and recognized on the basis of specified numbers of floral parts, among other features. Mutants with modified numbers of these flower structures also provide examples of significant changes at a level greater than the usual micromutational modifications in a species.

It is difficult to determine whether gene mutations of this magnitude of effect actually do initiate evolutionary novelty, even after stabilization with the accumulation of modifying genes that enforce the successful development of the new trait. Should we entertain the idea that the *vestigial* mutant of *Drosophila* may represent a macromutational system, which, in appropriate selective conditions, could spread through the population and become part of the genetic structure together with other alleles that become fixed in the genotype and stabilize the development and integration of the wingless condition? There are more than 150 known gene loci that contribute to normal wing development of *Drosophila*. Did these arise as modifiers subsequent to a macromutation leading to wing formation, or did wings gradually become established by a series of small mutational steps? It is difficult to choose between alternatives, and we should not necessarily discard the notion that rare macromutational events may initiate the relatively sudden appearance of new traits. There are few data to indicate that such macromutations may be successful; in fact, they often are quite detrimental to the mutant individual. But macromutation does seem less likely as an explanation, since any drastic alteration would have little selective advantage in a genetic background that had been tested and secured, step by step, over a long period of species evolution.

POLYPLOIDY

An increase in genome number may lead to instant speciation, since interbreeding would be ineffective between the new polyploid and its diploid ancestors because of the irregularities of chromosome pairing and distribution at meiosis. The high frequency of polyploidy among the flowering plants attests to the success of this phenomenon. Yet there is little evidence that polyploids lead to new forms in evolution, even though a polyploid species itself may, as is usually the case, be demonstrably more successful than its diploid relatives. The major observation is that all the principal adaptive traits for the group exist in the diploid species to begin with, and that the polyploids merely represent some variations on the basic species theme. Many experimental studies of polyploids have been reported, but since there can be no helpful fossil evidence for chromosome changes, our information comes only from living species. Still, the information is sufficiently clear to show that innovative evolution does not emanate from ploidy changes, even though these constitute an important element in higher plant evolution. Doubling of the chromosome number is rarely successful in sexually reproducing animals, because of sex chromosome imbalances, which occur on interbreeding with diploids. This process is considered to be of relatively little consequence in animal evolution in general.

ADAPTIVE DISPERSAL

In the case of adaptive dispersal we are dealing with a situation that does not concern genetic modifications; therefore, it is not a factor in innovative evolution. It is appropriate to discuss the phenomenon here simply to put it into a proper perspective relative to rapid spread of a new species of organism. In situations involving adaptive dispersal there is immediate success of fit genotypes introduced into new habitats. But the new habitats are essentially the same as those from which the populations came initially or as ones they had inhabited at some prior time in history. There are no changes in gene or genotype frequencies; hence adaptive dispersal is not a long-term evolutionary phenomenon per se.

Two examples will illustrate the main features. In 1859, 24 European rabbits were introduced into Australia. Three years later these animals had become a widespread pest, extending across much of the eastern and southern regions of that continent. The immediate success and consequent dispersal of the rabbits can be attributed to the similarities in climate and food sources in Europe and Australia and to the lack of competitors and predators in the new homeland. Despite numerous and continuing efforts, the spread of the rabbit populations still has not come under control in Australia.

Another substantiated example of adaptive dispersal concerns the horse, which was reintroduced into North America during the explorations by the Spaniards in the sixteenth century. The fossil record indicates that horses disappeared from North America during the Pleistocene ice ages, crossing the Bering Strait land bridge to Asia and continuing on to Europe. Successful evolution of the horse had occurred earlier in North America, and it continued in similar habitats elsewhere. Upon reintroduction to North America, large populations developed because the species had undergone its evolutionary development in this same region and it had been gone for an interim of only a few thousand years.

Transitional
Forms

The sudden appearance of successful innovations in the fossil record requires explanation. Wings appeared suddenly in insect species of the later Pennsylvanian epoch of the Paleozoic era: eyes appeared suddenly in Paleozoic arthropods; eyes appeared independently and as suddenly in vertebrates during the Ordovician period of the Paleozoic era; and so forth. How did these impressive improvements arise? Perhaps as pre-adaptive traits, which became stabilized by modifying gene action; perhaps by macromutations and the subsequent integration of modifying genes; perhaps by either or both of these mechanisms in conjunction with conventional micromutational steps; or perhaps only by conventional gradual fixation of small adaptive changes under the guidance of natural selection.

If small mutational steps were the principal factor, then why do we not find transitional forms in the fossil record for many groups of organisms? Any new mutational change must be integrated into a complex and harmonious genetic background, each component of which has undergone rigorous tests by selection over a long period of time. The successful persistence and spread of a new gene having a major effect almost certainly would occur together with the accumulation of suitable modifying genes, which reinforce the major change and lead to improved fitness of the entire changing genotype spectrum in the population. But the first mutational steps probably occurred in one or a few populations of the species. The first of the changing groups thus would tend to be small and to occupy a localized distributional area. Even if the mutational changes did lead to recognizably different features that were suitable for preservation, the probability for fossilization of these rare individuals would be quite low. We would expect that there would be very brief episodes of successful transition leading to the final stabilized new trait. The transiency of cumulative improvement steps would compound the problem of fossilization of rare specimens, and such intermediate forms would be unlikely to find their way into the fossil record. If there were some stabilized features of an intermediate character, such that these would spread and comprise sufficiently large populations over a broad range of geography, then such forms might be incorporated in the fossil record. But for those new forms that evolve rapidly toward an innovative mode of life there would be low probability for chance preservations.

Even for traits that are recognizably intermediate, we would expect the transitional populations to have been small, fleeting in existence, occupying diverse habitats, and thus unlikely to contribute specimens for such rare events as fossilizations. To compound the problem, if such a rare find were discovered it might not be possible to establish whether it was a transitional representative or merely an aberrant variant, not necessarily typical of the whole population. All together, it is indeed an unusual discovery when any transitional form is uncovered. For most innovations in evolution we would expect to find no trace of the sequences of change that led ultimately to the novelty in evolution.

A final comment is in order here even though the evidence is just now becoming available for analysis. Eukaryotes contain substantial amounts of redundant DNA, amounting to 15 to 80 percent of the genome. Some of these sequences are repeated up to 1 million times in a chromosome complement. Several authors have made the suggestion that the repeat sequences of nucleotides may comprise groups of regulatory genes, which monitor the action of the unique nucleotide sequences of structural genes that are translated into particular proteins. If this can be shown to be the situation in complex life forms, then we can extrapolate the system to help explain the speed with which novelties become established in evolution. The speculation is that new adaptive mutant genes would arise by the conventional mechanisms and produce conventional micromutational changes. But such mutant loci would become stabilized very rapidly and spread quickly through the population if they were accompanied by stretches of redundant DNA in the same genotypes, exerting regulatory control over protein translation from structural gene information. The significant element in the evolution of novel complex life may therefore reside in the controlling DNA components that modulate the pathways of developmental processes; and these processes may require relatively few structural genes specifying relatively few major proteins. Genetic information specifying proteins may be present in different quantities in simple and complex life, but the quantity and quality of redundant DNA may provide the key to the progressive evolution of increasingly complex life forms and novel modes of exploiting a life style.

Some Suggested Readings

Dobzhansky, Th., *Genetics of the Evolutionary Process*. New York, Columbia University Press, 1970, pp. 267–390, 407–414.

Fooden, J., "Breakup of Pangaea and isolation of relict mammals in Australia, South America, and Madagascar." *Science* 175 (1972), 894.

Mayr, E., *Animal Species and Evolution*. Cambridge, Massachusetts, Harvard University Press, 1963.

Mayr, E., "Evolution at the species level," *Ideas in Evolution and Behavior* (ed., J. A. Moore). Garden City, New York, Natural History Press, 1970, pp. 315–325.

Stebbins, G. L., *The Basis of Progressive Evolution*. Chapel Hill, University of North Carolina Press, 1969.

CHAPTER 12

VERTEBRATE EVOLUTION

The Animal Kingdom is subdivided into fourteen phyla (singular, phylum) of multicellular, multinucleate forms (metazoans) and two phyla (Protozoa and Porifera, or sponges) that are considered to be Protista in the modern five-kingdom scheme of classification (see Fig. 5.7, p. 89) or Animalia in the older conventional formats (see Appendix A). The colloquial terms for the metazoan animals are invertebrates and vertebrates. The invertebrates include most of the known species of organisms (perhaps three-fourths are in the insect group alone), such as worms, molluscs, starfish, leeches, insects, and many other types, all of which lack the internal organization of the vertebrates, which we will mention in a moment. Vertebrates comprise a subphylum of the phylum Chordata. The other three chordate subphyla include such few and inconspicuous species that the vertebrates have been emphasized in the dichotomous view of the Animal Kingdom. The most probable invertebrate group to have provided the chordate ancestor is the phylum Echinodermata, to which the starfish, sea urchins, and related marine forms belong.

Chordates and invertebrates share various common features of organization and development, including bilateral symmetry, the differentiation of three primary germ layers in the embryo, and a coelom or true body cavity among the most significant traits. Three unique features distinguish the chordates from all other animal phyla (Fig. 12.1). Each of these features occurs at least at some stage of development of each chordate, although

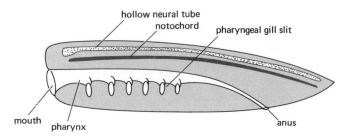

Figure 12.1   Schematic drawing showing some major chordate traits.

they may not persist throughout the life of each kind of organism. There are (1) a dorsally-located *notochord,* a pliant connective tissue, which appears in the embryo stage of all chordates but is replaced by the backbone in all adult vertebrates; (2) a hollow dorsal *neural tube* located just above the notochord, which represents the central nervous system; and (3) *pharyngeal gill slits,* which function as respiratory organs throughout life in the simple aquatic forms but which have evolved to new systems of structure and function in the terrestrial chordates. These are transient nonfunctional embryonic structures in terrestrial forms and become modified into systems such as the middle ear, among others. The four subphyla of the Chordata include the following:

1. Hemichordata (acorn worms), of which about 100 marine species are still living.
2. Urochordata (tunicates and sea squirts), which include about 2000 marine species today.
3. Cephalochordata (*Amphioxus* and lancelets), which are represented by about 30 extant marine species.
4. Vertebrata, which include the dominant forms of complex animal life on the planet, grouped into two superclasses, each of which is subdivided into four classes of organisms:
   A. Superclass Pisces (aquatic forms)
      a. Class Agnatha (jawless forms such lampreys and hagfishes)
      b. Class Placodermi (earliest jawed fishes, now extinct)
      c. Class Chondrichthyes (groups with cartilaginous skeleton, including sharks, skates, and rays)
      d. Class Osteichthyes (all true bony fishes, such as trout, salmon, carp, and many others)
   B. Superclass Tetrapoda (four-footed terrestrial forms)
      a. Class Amphibia (such as frogs, toads, and salamanders)
      b. Class Reptilia (such as turtles, snakes, lizards, and crocodiles)
      c. Class Aves (all birds)
      d. Class Mammalia (including the three major subclasses of warmblooded species, which have skin covered by hair and suckle their young on milk secreted by mammary glands of the female)
         (1) Subclass Prototheria (egg-laying species, such as platypus and spiny anteater)
         (2) Subclass Metatheria (marsupials such as kangaroo)
         (3) Subclass Eutheria (with placental development system for the embryo; many orders, including the Primates)

All vertebrates have a vertebral column, which completely replaces the embryonic notochord in the more advanced groups. The notochord persists throughout life in the less evolved groups, running through the center of each vertebra in the column. Most vertebrates have, among other features, the following:

1. a brain enclosed within a cranium, the brain being a differentiated region of the neural tube
2. two pairs of jointed appendages, as limbs or fins
3. a closed blood-vascular system characterized by the presence of red blood cells and a pumping heart
4. dorsally located kidneys
5. paired ears and eyes
6. gill slits, which are permanent in the lower forms and transitory in the others
7. a single pair of gonads in the separate sexes
8. a large coelom or body cavity
9. a portion of the body projecting rearward as a tail

Evolutionary Relationships

The earliest known vertebrates are the fossil jawless fishes (Agnatha) from Paleozoic deposits of the Ordovician period. No vertebrates have been found from the earlier Paleozoic period, the Cambrian, although many invertebrate forms abounded in the Cambrian seas. Extinct groups, such as those of the order Ostracodermi, are found no later than the Devonian period, about 350 million years ago, but living representatives of the agnathans include lampreys and hagfishes. The jawed fishes first occured in the Silurian period, and all Placodermi became extinct by the end of the Devonian period along with the jawless ostracoderms and many other groups at this time of widespread mass extinction. The placoderms probably were descended from the jawless agnathan fishes, and the jawed groups in turn are believed to be ancestral to all later fishes (see Fig. 10.1).

The groups of sharks probably were descended from early Devonian placoderms (arthrodires) and themselves were abundant in the Devonian seas. The sharks have always been marine forms, as are the more highly evolved skates and rays among the Chondrichthyes. These groups lack lungs and swim bladders, unlike the true bony fishes. At one time it was believed that the Chondrichthyes were directly ancestral to all vertebrates, but additional evidence, together with the fossil record, clearly indicates that they are a divergent line descended from bony placoderms. Unlike the bony fishes in the other three classes of Pisces, sharks and their relatives retain their cartilaginous skeleton throughout life rather than replacing that material with bone during development. This case provides an example of the reinterpretation of assumed evolutionary sequences on examination of more definitive evidence from comparative anatomy and embryology, along with the time coordinates from the fossil record.

The true bony fishes (Osteichthyes) probably originated as marine forms very late in the Silurian or very early in the Devonian period. Modern bony fishes now occupy both marine and freshwater habitats. The ray-finned fishes are considered to be ancestral to present-day bony fishes. The earliest forms had a pair of lungs as well as gills. During evolution there was a reduction to one lung, which still is functional in a few kinds of modern forms but which has been modified to a swim bladder in teleost fishes. The swim bladder in teleosts generally has a buoyancy function, but sometimes it acts as a producer or receptor of sounds.

The important observation is that lungs evolved before swim bladder organs in the bony fishes. The lobe-finned bony fishes are acknowledged to be ancestral to existing lungfish and to all terrestrial vertebrates (Fig. 12.2). Although the ancient lobe-fins were freshwater types and capable of breathing air (nostrils also were present), the few remaining lungfish are marine species in southern waters. In these marine species the lung functions as an organ of fat storage and provides aid in buoyancy, rather than acting as a respiratory structure. The ancient air-breathing lobe-fins are classified as members of the order Crossopterygii and were believed to be extinct until very recently. Beginning in 1939 living coelocanths were reported caught by local fishermen off the coasts of South America and Madagascar. These living crossopterygians are believed to be a persistent side branch of the ancient group rather than relics of the ancient stock itself. The fin bones of coelocanths and other crossopterygians are directly comparable to amphibian limbs in skeletal anatomy. From evidence such as this and from the fossil record it seems most probable that terrestrial vertebrates were descended from air-breathing lobe-fin ancestors some time toward the end of the Devonian period, more than 350 million years ago.

The earliest known fossil amphibian (*Ichthyostega*) was found in eastern Greenland deposits. Although they resembled fishes in many respects these creatures clearly were four-footed and therefore members of the Tetrapoda. These early amphibians, called labyrinthodonts, appear in the fossil record long before extensive evolution of the sharks and the bony fishes, which are found in later deposits in greater numbers and diversity than their mid-Paleozoic relatives. The labyrinthodont amphibians were extinct by the end of the Triassic segment of the Mesozoic era, having endured for almost 200 million years. There are only three living orders of amphibians: the tailed newts and salamanders, the limbless wormlike forms, and the frogs and toads, which are tailless in the adult stage of development. Amphibians are varyingly terrestrial, but all require an aquatic medium for reproduction because fertilization of eggs by sperm and the subsequent development of the larvae all take place in water. Since reptiles appear in the fossil record long before there is any evidence of the existence of the frog and sala-

Figure 12.2 Painting showing lobe-finned Crossopterygian fishes of the Devonian period. (Photograph courtesy American Museum of Natural History, New York.)

mander groups, it is clear that the terrestrial habit in reptiles evolved independently and earlier than a more sustained terrestrialism in modern amphibians. In other words, reptile land existence stems from a set of genetic modifications different from those that permitted an enhanced life on the land in present-day Amphibia.

Members of the class Reptilia were the first true terrestrial vertebrates, being totally independent of water for all phases of their life cycle. There is no larval stage, as in amphibians; the young resemble the adults at birth. The primary evolutionary modifications involved changes in reproduction, development, respiration, circulation, and anatomy, all of which permitted a terrestrial existence. The evolution of systems for internal fertilization of eggs by sperm and for the protected development of the embryo within the egg were especially critical innovations. The hard covering of the egg prevented drying out and insulated the embryo against extreme temperature fluctuations, and the contents of the egg provided a food supply for the embryo during development. Although it is not altogether certain, it is believed that reptiles may have been descended from such labyrinthodont amphibian types as *Seymouria,* at least 300 million years ago in the early part of the Pennsylvanian epoch. The first reptiles may have appeared even earlier; we have an abundant fossil record for the pelycosaur reptile group in the mid-Pennsylvanian, a rather short time after the presumed time of origin for the entire class Reptilia. In addition to the pelycosaurs and other groups of reptiles, the therapsids (mammallike reptiles) also enjoyed rapid and abundant diversification during the Permian and Triassic periods. Pelycosaurs became extinct at the end of the Paleozoic era, but therapsids continued to evolve during the first part of the Mesozoic, becoming extinct early in the Jurassic period. The therapsids (Fig. 12.3) are ancestral to the mammals, which first appeared at the end of the Triassic period, about 200 million years ago, but other reptile stocks diverged into numerous adaptive lines during the Age of Reptiles. Modern derivatives of these ancient reptilian lineages include the turtle, snake, lizard, and crocodile groups.

The Mesozoic dinosaurs were descendants of an archosaur line, which remains today only in the vestiges of the crocodile group. Thecodonts were lizardlike reptiles of the early Triassic, which ran on their hind legs; they diverged to produce many of the running and arboreal types as well as the very large phytosaurs of the late Triassic, these resembling crocodiles in many ways. In addition to siring the order Crocodilia, thecodonts produced other divergent lines, which continued to evolve into the pterosaurs, or flying reptiles, and the new vertebrate class of birds (Aves). The fossil evidence indicates that birds and pterosaurs evolved independently from different groups of arboreal thecodonts.

There are only four living orders of reptiles: (1) the turtles, terrapins, and tortoises, which represent the most primitive group; (2) the crocodiles and alligators, which are quite advanced forms that have returned to the water in their subsequent evolution; (3) the lizards and snakes, the most abundant of the living reptiles, including about 4000 species; and (4) a group that has only one living member (*Sphenodon*), which is near extinction and today is found only in New Zealand.

Figure 12.3   Painting showing a restoration of the therapsid reptile *Lycaenops*, a member of the group that gave rise to the mammals. (Photograph courtesy American Museum of Natural History, New York.)

The earliest fossil bird was the long-tailed, toothed *Archaeopteryx* (Fig. 12.4), which was discovered in mid-Jurassic deposits in central Europe. The birds were very much like reptiles in many features, including the feathers, which are modified reptilian scales. Although many of the reptile groups, including dinosaurs, icthyosaurs, monosaurs, pterosaurs, and others, became extinct at the end of the Mesozoic era, more than 60 million years ago, birds have become increasingly abundant and diversified during the past 160 million years. There was a relatively rapid evolution of shorter-tailed toothed birds until near the end of the Cretaceous period, when the first modern (toothless) birds appeared. In addition to all the modern orders of birds, various gigantic species also were produced during the Cenozoic era. These included the flightless moas of New Zealand, elephant

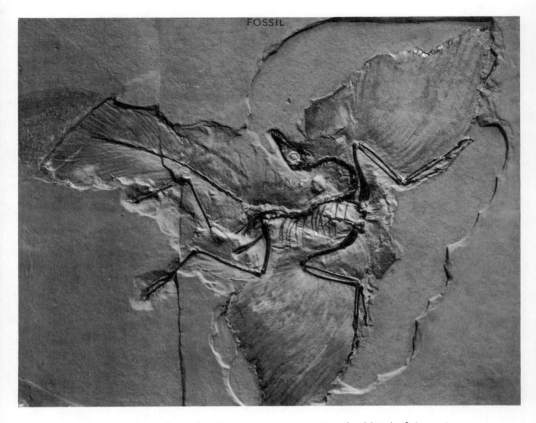

FOSSIL

Figure 12.4   Cast of the fossil *Archaeopteryx*, a toothed bird of Jurassic times showing the long tail of the primitive bird groups. (Photograph courtesy American Museum of Natural History, New York.)

birds of Madagascar, and the largest species of bird ever known, the flying sea bird *Osteodontornis*.

Many of the evolutionary developments that characterize the birds are found also in mammals, but these traits arose independently in the two classes and are based on entirely different genetic modifications. The ancestry of the two classes is different, as we have mentioned already, although in both cases it is reptilian, birds having evolved from a thecodont line and mammals from a therapsid lineage. Among the significant avian developments are the internal control of body temperature, which permits high levels of activity based on high metabolic rates and the rapid consumption of oxygen, and a highly developed central nervous system (especially cerebellum, optic lobes, and cerebral hemispheres) in correlation with control of motor activities, the senses of sight and equilibrium, and complex behavior patterns.

Mammal
Evolution

Mammalia are descended from therapsids, which in turn were products of evolution from early Permian carnivores of pelycosaur lineage. Most of the therapsids became extinct in the Triassic, but a few persisted into the very early Jurassic period. During their existence, somewhat less than 100 million years, there were carnivorous and herbivorous therapsid lines distributed over the whole world. The transition from more reptilelike, to more mammallike, and finally to mammals occurred near the end of the Triassic period. It is controversial whether mammals all are descended from a single therapsid lineage or from several different groups of these mammal-like reptiles. The difficulty resides in the fact that many of the distinctive mammalian traits (Fig. 12.5) are related to physiology and reproduction, features of therapsid evolution about which little is known, the principal

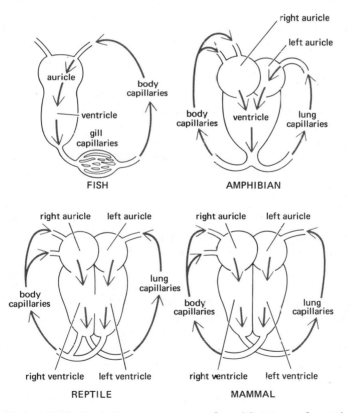

Figure 12.5   Evolutionary sequence of modifications of vertebrate heart and circulatory system, as determined from studies of living forms.

fossil evidence available for interpretation consisting of skeletal remains (Fig. 12.6).

One of the skeletal changes considered to be of profound importance, and therefore primary evidence for constructing evolutionary sequences, involved the bones of the lower-jaw–middle-ear complex (Fig. 12.7). The lower jaw in mammals consists of a single bone (the dentary) on each side, which is articulated with the skull differently from the way the reptile jaw is. There are several bones comprising the reptilian lower jaw, and those same skull bones that are involved in articulation in reptiles have been greatly modified in mammals, now constituting the sound-transmitting apparatus of the mammalian middle ear (Fig. 12.8). This particular sequence of modifications provides one of the most widely used lines of evidence for interpreting the lineages, the timing, and the recognition of the first mammals in 200-million-year-old Triassic deposits.

There are three living subclasses of mammals: (1) the Prototheria or simple egg-laying monotremes, of which the duckbill, or platypus, and the echidna, or spiny anteater, of Australia are the only extant types; (2) the Metatheria or pouched mammals, which include fossil and living marsupial species such as the opossum of the Americas and the kangaroos, koalas, wombats, and other Australian fauna; and (3) the Eutheria or placental mammals, which are the dominant forms on the planet today and are subdivided into ten distinct living orders of which the Primates are one.

There is no fossil record for the egg-laying monotremes, but the living forms have retained some reptilian reproductive and skeletal features while having undergone considerable evolutionary modifications of the skull, jaws, and teeth. The platypus has only hard pads instead of teeth in the adults, and the spiny anteater has no teeth at all. In both of these types the skull and jaw are modified to form an elongated beaklike or snout projection, used by the platypus for burrowing for worms and grubs in the mud bottoms of streams in which it lives, and by the anteater to forage in ant-hills in its deep forest habitat. Despite such specializations to a circumscribed existence, monotremes are extremely primitive mammalian forms. They reproduce by laying eggs, which are hatched in burrows, and the young suckle on oozing milk secreted from modified sweat glands, which are homologous with the mammae or breasts of higher forms. From a number of skeletal and anatomical features closely resembling an ancient mammalian lineage known from scanty fossil remains, some consider it quite possible that the modern Prototheria were derived by a different line of descent, perhaps as early as the late Jurassic or early Cretaceous periods of the Mesozoic. Aside from that, monotreme fossils are not known from deposits older than the Pleistocene epoch of quite recent times.

The marsupial mammals diverged during the mid-Cretaceous period and were abundant in South America and in Australia during the approximately 60 million years of the Cenozoic era. The fossil record is quite fragmen-

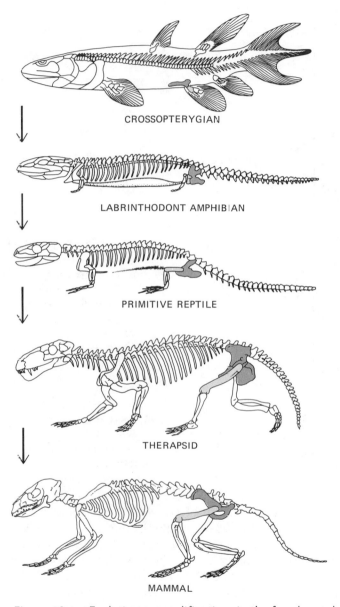

CROSSOPTERYGIAN

LABRINTHODONT AMPHIBIAN

PRIMITIVE REPTILE

THERAPSID

MAMMAL

Figure 12.6 Evolutionary modifications in the four-legged skeletal anatomy of vertebrates. (Adapted from *Biology: Observation and Concept* by Case and Stiers, 1971, Fig. 15-11. Copyright © 1971 by The Macmillan Company. Reprinted with permission of The Macmillan Company.)

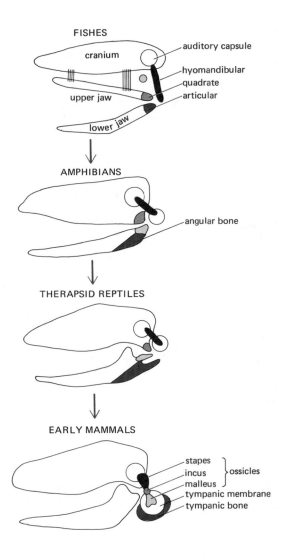

FISHES

cranium
auditory capsule
hyomandibular
quadrate
articular
upper jaw
lower jaw

AMPHIBIANS

angular bone

THERAPSID REPTILES

EARLY MAMMALS

stapes
incus } ossicles
malleus
tympanic membrane
tympanic bone

Figure 12.7 Evolution of the bony structure of the vertebrate ear, showing modifications leading to the uniquely mammalian middle-ear— lower-jaw complex. (Adapted from *Biology: A Functional Approach* by Roberts, First Edition, 1971, p. 567. Copyright © 1971 by The Ronald Press Company, New York. Reprinted with permission of the publisher.)

tary for marsupial species in other parts of the world; marsupials did not fare too well elsewhere if they competed with the more successful evolving placental mammals. From reconstructions of continental migrations during critical times of mammalian evolution over the past 200 million years, there have been suggestions made to explain the present disjunct distribution of monotremes and marsupials relative to the more highly evolved placental groups. The peculiar distribution of primitive mammalian groups today is especially evident in the presence of monotremes and marsupials in Australia, marsupials and primitive placentals in South America, and primitive placentals but no monotremes or marsupials on the island of

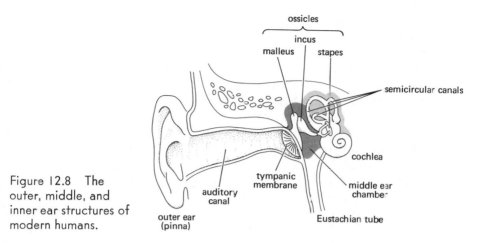

Figure 12.8 The outer, middle, and inner ear structures of modern humans.

Madagascar. If there was the supercontinent of Pangaea during the late Triassic, as we believe, then we would expect that the evolving early mammals would have included the ancient monotreme and marsupial types. These forms would have constituted a world fauna because land connections were available everywhere in Pangaea. As Pangaea fragmented into Gondwana and Laurasia, the monotremes and marsupials became isolated in Antarctica–Australia before the appearance of placental mammals in all parts of the world of that time. The two more primitive groups continued to evolve in East Gondwana, but this region was inaccessible to the fauna, including primitive placentals, isolated on the land mass of West Gondwana and Laurasia. Primitive monotremes became extinct as the evolving marsupial and placental species increased and diversified in regions other than Antarctica–Australia (East Gondwana). Continued continental fragmentation and drift led to the separation of South America from Africa and to the isolation of the South American marsupial and primitive placental fauna. The bulk of the placental evolution continued to occur in the remainder of the world, and extinction of the marsupial groups took place as the more efficient placental species became predominant. Marsupial forms declined in South America as the advanced placental species inhabited an increasing proportion of the living zones and as migrations of placentals occurred between North and South America with the appearance of the Central American land link later in the Cenozoic era. The only remaining marsupial of the Americas is the opossum, considered to be a "living fossil" because it is so little changed from its relatives of many millions of years ago in the New World.

Marsupial species bear their young alive, but these offspring continue a substantial part of their embryonic development in the mother's pouch, having crawled there from her uterus after a very brief gestation. The successful young attach themselves to the teats in the pouch and remain so attached for a long time as milk literally is pumped into them by the mother's efforts. Young marsupials continue to spend part of their time in the mother's pouch, but they remain outside on their own for increasingly longer periods as they come to assume more independence. Such features of reproduction and behavior in the care of the young are based on observations of modern metatherian species, not on fossil remains. Skeletal characteristics that serve to identify fossil marsupials include a very small braincase and a number of features of the jaws, teeth, and skull. The American opossum, which has changed slightly from its Cretaceous ancestor, provides a wealth of information that can provide greater insights into the habits and features of Mesozoic marsupial mammals.

Pliocene extinctions of South American marsupials, except for the opossums, probably were due to the competition with the more advanced placentals, which were well diversified by this time. In Australia, unlike the Americas, there were no placental competitors and the marsupials enjoyed a relatively extensive evolution there. Most of the arboreal and herbivorous marsupial species are still living. The introduction of placental mammals when humans reached Australia, perhaps 15,000 to 20,000 years ago, led to the rapid demise of some indigenous marsupials, which did not survive the new competition. The placental dingo dog essentially replaced the marsupial carnivore equivalent, the Tasmanian wolf, and similar extinction patterns undoubtedly occurred in other parts of the world in earlier times when placental species replaced marsupials inhabiting equivalent ecological zones.

The placental mammals probably had diverged from the marsupials in the western hemisphere at least 100 million years ago, during the Cretaceous period. The fossil evidence from western North American deposits reveals that the marsupial opossums and the most primitive of the placental mammals, the insectivores, were strikingly similar in jaw and tooth patterns. Although there were diverging groups of placental mammals during the late Cretaceous, no modern orders of Eutheria were present at that time. The primitive placental groups included a number of lines that had diverged from the archaic insectivores, including carnivores, ungulatelike types, and primatelike types. The dinosaurs were the predominant terrestrial animals during the time the mammals were evolving in the Cretaceous period. By the end of that period the placental mammals had radiated extensively over most of the world, and the principal period of evolutionary development occurred during the Cenozoic era, after mass extinction of many reptile groups at the end of the Mesozoic. Primitive primates, the order to which humans belong, were present at the beginning of the Cenozoic era, but no evidence has been found to indicate that primates existed earlier than this time.

The placental innovations were of many kinds, but some were significant improvements over earlier and more primitive mammalian forms. The embryo of the placental mammal is nurtured in utero until a longer developmental term is completed, in contrast with marsupial embryos, which move from the internal uterus to the external pouch very shortly after gestation has begun. The placenta is an organ composed in part of maternal tissues and in part of embryonic tissues; it provides nourishment and circulatory connections with the mother for the developing fetus. The dependence on parents continues for a variable period of time after birth of the placental mammal infant, some being unable to manage at all for some time after birth, others being capable of a high level of independence at birth. Such animals as horses and cattle produce young that have fur and sight and hearing plus a level of motor coordination that permits the newborn to walk almost immediately. The newborn of other animals, such as rodents, are naked, blind, and deaf. The offspring undergo a period of postnatal development under parental care before they are capable of some independent activity. In addition to these improvements in the placental species, one of the most significant advantages undoubtedly came from modifications that led to a higher level of intelligence than is typical of marsupial species. The enhanced mental capacity of the placental mammals probably accounted for their triumph over marsupials and their predominance in the Cenozoic fauna of the principal areas of the world.

Primate
Evolution

The human species is amenable to classification in the same way as any other life form, such schemes serving to indicate relationships and ancestry. A proper understanding of the evolutionary history of the human species can be mirrored in the assignments made by systematists, because classification is based on genetic relationships, hence on evolutionary relationships. Since humans have nucleated cells, they are eukaryotes, and they belong to the Animal Kingdom by virtue of possessing a multicellular, tissue-forming organization that contains cells lacking chlorophyll and rigid cell walls. We belong to the phylum Chordata because we have a notochord (during embryonic life), a neural tube (differentiated into a complex central nervous system), and pharyngeal gill slits (while in the embryonic state). Specifically we are vertebrates among the chordates, and members of the tetrapod class Mammalia because humans suckle their young, have skin covered with hair, and show the variety of other traits typical of mammals. Among the mammals we are eutherians because the fetus is associated with a placenta during its gestation (the placenta is known colloquially as the "afterbirth"). Once we have run down our relationships to this point in the classification scheme, we come to the point of deciding the particular order of eutherians to which we belong. Clearly we are primates, not bats or whales or horses or elephants or any of the other eutherian types. Along with other primates, humans have a number of distinctive features. But we also differ from our fellow primates, such as lemurs and monkeys and gibbons and gorillas. Thus the formal systematics permit us to work right through to identification of the species, since each life form has a unique set of traits that distinguishes it from all others of past and present times. After viewing the formal classification of the human species, we will proceed to examine our closer relatives and thus gain further understanding of our evolutionary relations and our history.

Humans are eukaryotes of the Animal Kingdom, and also:

Phylum Chordata
 Subphylum Vertebrata
  Class Mammalia
   Subclass Eutheria (placentals)
    Order Primates
     Suborder Anthropoidea (one other: Prosimii)
      Infraorder Catarrhina (one other: Platyrrhina)
       Superfamily Hominoidea (includes gibbons, great apes, and humans)
        Family Hominidae (two others: Hylobatidae, the gibbons; and Pongidae, the great apes)
         Subfamily Homininae (one other: extinct Australopithecinae)
          Genus Homo (no other recognized, living or extinct)
           Species Homo sapiens (at least two others, both extinct)

Although there is no single distinguishing feature common to all members of the order Primates, they all share a characteristic *lack* of specializations. A general set of "prevailing tendencies" provides the basis for identifying primates, whether of living or fossil forms. The limb structure is generalized (see Fig. 10.5) and associated with free mobility (flexion–convergence) of the digits and especially of the thumb and toe. Sharp, compressed claws have been replaced by flattened nails, with sensitive digit-tip pads correlated with the occurrence of nails. The limbs are elongated and have five-digited hands and feet. There has been an elaboration of the visual powers and a corresponding reduction in the olfactory apparatus and thus in the sense of smell. There is a shortened snout or muzzle, the progressive development of a large and complicated brain, and the preservation of a simple pattern of the cusps and surfaces of the molar teeth.

There are variations, too, among the primates. Tree shrews have 20 claws, but the brain is slightly enlarged and the eyes are just beginning to be directed forward as compared with the insectivores to which they had been assigned until very recently by the systematists. Some primate species have both nails and claws, such as the New World marmoset monkeys, which have 18 clawed digits but flattened nails on the great toes. The aye-aye of

Madagascar not only has nails on its great toes and claws on the other 18 digits, but is even more unusual in having chisellike, gnawing incisor teeth similar to those of rodents and quite unlike primate incisors. Modifications of general primate features in humans include an especially large and complex brain; considerable modification of hind limbs, related to upright posture and bipedal locomotion; and changes in the feet such that the great toe no longer is opposable and there is little free mobility of all the toes, thus limiting grasping abilities and possibilities.

The living primates include the primitive prosimians (Fig. 12.9), remnant populations of the earlier widespread group, and the various anthropoid groups. Among the prosimians the most primitive are the Oriental tree

Figure 12.9 Painting of one of the lemur species of Madagascar, a living prosimian. (Photograph courtesy American Museum of Natural History, New York.)

shrews, which are about the size of a mouse. They are arboreal, insect-eating, diurnal in habit, have 20 clawed digits, and occur in a variety of species, now restricted to regions of southeast Asia. They probably are very similar to the ancestral form, which diverged from an insectivore lineage just before or at the onset of the Cenozoic era. There are two general groups of lemurs, restricted in distribution today but worldwide in early Cenozoic times. The true lemurs occur in a variety of species on the island of Madagascar off the southeast coast of Africa, in the present Malagasy Republic. These lemurs are in danger of extinction today as a result of human activities and the encroachment of civilization on the forest habitats. The aye-aye, which has Asian affinities, is an unusual lemur found on Madagascar, as mentioned above. A second lemuroid group is represented by remnant populations on the African continent, including the galago, potto, and bush baby, and by the lorises of India on the Asian continent. These lemuroids are nocturnal, arboreal, and insectivorous. The third type of prosimian exists today as the single genus *Tarsius*, restricted to the Philippines and to the Indonesian islands of Borneo and the Celebes. These also are nocturnal, arboreal, and insectivorous in habit, and possess a mixture of features that may be considered primitive and advanced for a prosimian group. Tarsiers, which are about the size of a six-week-old kitten, can leap at least six feet with great accuracy because their hind limb development is rather unusual.

There are two groups of living anthropoids, the platyrrhine New World monkeys and marmosets and the catarrhine Old World monkeys, gibbons, great apes, and humans. Platyrrhine species have broad, flat nose cartilages, which lead to laterally disposed nostrils, whereas catarrhines have parallel nostrils because of the narrow partition formed by the compressed nose cartilages. The distinction is not always particularly striking. The New World platyrrhines have a prehensile tail, which functions virtually as a "fifth hand," aiding in movement through the trees and in stability by virtue of its grasping capacity. The catarrhines are divided variously into groups, but we may consider the Old World monkeys as separate from the hominoid aggregate of gibbons, great apes, and humans. Among the Old World monkeys are many species, such as baboons and macaques, which are variously arboreal or terrestrial or inhabit both kinds of living zone in their daily cycle of activities. The gibbons are quite distinct and can be separated into the family Hylobatidae (Fig. 12.10). These are the familiar free-swinging small apes seen in almost every zoo. There are three types of great ape, members of the family Pongidae: the arboreal orangutan of Indonesia and the chimpanzees and gorillas of the African continent, which are more terrestrial but retire to the trees at night. We reside alone in the family Hominidae, but there are fossil relatives, which are included in this group along with modern *Homo sapiens.* The anthropoids, or higher primates as distinguished from the prosimians, all have well-developed stereoscopic

Figure 12.10   A gibbon, one of the hominoid primates. Note the relatively long arms of this brachiating species.

vision and color perception, and possess acute vision. These characteristics permit the successful daytime activities of the group in contrast with the predominantly nocturnal prosimian species. Anthropoids generally are larger than prosimians, but a more significant distinction is the occurrence of an increasingly larger brain with an especially well-developed cerebral cortex. There is a progressive increase in the size of the cerebrum, which covers the other parts of the brain, and in humans the cerebrum substantially overlies all the rest of the brain.

THE FOSSIL RECORD OF THE PRIMATES

Although mammals date from the Triassic period, the first primates do not appear in the fossil record until the beginning of the Cenozoic era, more than 60 million years ago. These early primates were small, arboreal, insect-eating creatures, which were clearly descended from an insectivore lineage (see Fig. 10.7). In fact, the transition is difficult to discern in the earlier fossils, just as it is for such archaic types as the living tree shrews. The principal fossil materials consist of skeletal remains; therefore these provide the basis for interpreting evolutionary trends in the group.

From changes in the cranium to larger size and more globular shape, it can be inferred quite properly that the brain enlarged and became more complex. Full stereoscopic vision became incorporated, as can be seen from the progressive rotation forward of the eye orbits and the enclosure of the orbits by a bony ring. In the higher primates the orbits are completely shut off from behind by the bony ring. The jaws were reduced as the bony arch of the cheek (zygomatic arch) became attenuated, thus also indicating less emphasized masticator muscles, since these are attached to the zygomatic arch. The reduction in the sense of smell can be detected in fossils, both from the reduced size of the bony apparatus of the nose and from the progressive recession and retraction of the face to a position below rather than in front of the cranium proper. The articulation of the whole skull with the neck region of the backbone improved progressively, according to the fossil series. The foramen magnum (the hole in the skull through which the spinal cord passes in continuity with the brain) has been displaced forward with a concomitant forward displacement of the occipital condyles, bony processes that permit the articulation of skull and upper part of the backbone (Fig. 12.11). In humans the displacements of these structures have culminated in positions that permit the head to sit directly above the backbone, thus complementing other anatomical changes associated with continuous upright posture and related bipedal locomotion (Fig. 12.12).

Changes in dentition are indicative of altered food habits and defense postures. The ancestral mammalian stock showed the half-jaw dental formula of incisor–canine–premolar–molar teeth to be

$$\frac{3.1.4.3}{3.1.4.3} \times 2 = 44$$

Generalized dentition still characterizes most of the primates, but various changes have occurred, both in numbers of the different kinds of teeth and in specific features of the premolars and molars. For example, human premolar teeth have two cusps rather than one and the molars have four to five cusps as compared with three in ancestral types. The dentition formula for humans is

$$\frac{2.1.2.3}{2.1.2.3} \times 2 = 32$$

Incisor teeth generally have a shearing function, canines are tearing teeth, and premolars and molars are teeth utilized in grinding food. The variations in dentition can be related easily to the kinds of food that comprise the species diet. The omnivores, which eat both meat and vegetal foods, as do humans, have the full complement of varied functional kinds of teeth, whereas strictly vegetarian animals usually have conspicuously emphasized grinding teeth and canines that are reduced and modified. Carnivores have greatly enlarged canine or tearing teeth and reduced grinding types. Large and well-developed canine teeth, however, may be related to social behavior rather than to food habits. In such cases it is common to find that the size of the canine teeth is sex-influenced, males having very large canines while females of the species show considerably smaller teeth of this type.

Prosimians were widespread during 30 million years of Cenozoic times and were considerably diversified by the end of the Eocene epoch. Various

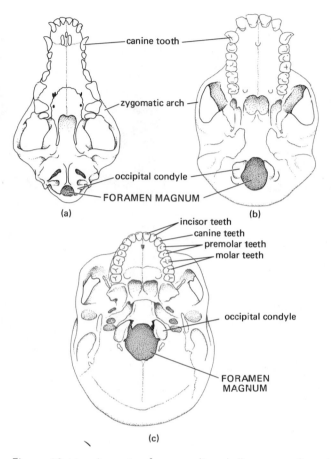

Figure 12.11   Aspects of mammalian skulls, as seen from below after removal of the lower jaw. The foramen magnum is located farthest forward in humans, and the shape of the upper palate assumes a more foreshortened and rounder aspect. Skull of (a) a dog, (b) a chimpanzee, and (c) a human. (Reprinted by permission of the University of Chicago Press from *History of the Primates* by W. E. L. Clark, 1965, Fig. 3, p. 22, 5th edition. Copyright © 1965 by the Trustees of the British Museum of Natural History. All rights reserved.)

prosimian forms occurred in Europe and America by the end of the Eocene but disappeared from these regions during the Oligocene epoch. The disappearance of prosimians from these areas coincided with the presence of the first anthropoids in the Oligocene, about 40 million years ago. Prosimians continued to migrate in other parts of the world, however, and the fossil lemurs of late Oligocene to early Miocene times were very much

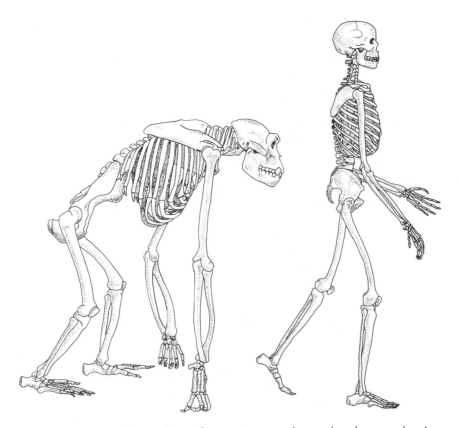

Figure 12.12 Comparison of a great ape and a modern human, showing similarities and differences typical of the Pongidae and Hominidae, respectively.

like the modern specialized lemurs of Africa. Lemurs had reached Asia by Pliocene times, and at least one group diversified on the island of Madagascar to produce a number of extraordinary Pleistocene species. Some of the Madagascar species superficially resemble monkeys. One large species still may have been present in 1658. A French explorer described an animal about the size of a two-year-old calf with a rounded head, humanlike face, fore and hind feet like a monkey's, and ears like those of a man. This lemur species (*Megaloadapis*) has not been reported since that time, except as fossils.

The fossil record is too fragmentary to determine the lineage among prosimians, and it is considered uncertain whether tree shrew types were ancestral to both lemurs and tarsiers or only to the lemurs. If the latter is true, it is possible that tarsiers were descended from some lemur or lemur-like species. All three groups of prosimians were present in the Eocene epoch, and the only certainty is that the earliest forms were very much like the modern tree shrews. The rapid spread and diversification during the Eocene indicate an adaptive radiation phenomenon for prosimian primates. With improved vision and the capability for swift travel through the trees, as well as increasing intelligence, the prosimians were admirably adapted to an arboreal existence, and they exploited this living zone in a number of different ways. The withdrawal of prosimians from Europe and North America during the Oligocene epoch was paralleled by the appearance of the first of the anthropoid higher primates. The tarsier group seems the most probable ancestor of the first anthropoids, which probably were much like monkeys. Fossil anthropoids from Oligocene deposits are rather generalized forms showing definite traces of tarsioid ancestry.

EVOLUTION OF THE HOMINOIDS

Most of the fossil information comes from Old World rather than New World species. The record is especially poor for the monkeys, but it does show that the Old World and New World monkey lines already had diverged from hominoids by the beginning of the Oligocene epoch. Whether hominoids diverged from an Old World monkey ancestry or directly from tarsioid lineage is uncertain. If the latter is true, then Old World monkeys may represent a separate evolutionary divergence from a common form that was ancestral to several anthropoid lines, including monkeys and hominoids. Oligocene fossils further show that gibbon types were distinguishable from the more generalized hominoids that ultimately produced the great apes and the species belonging to the human family. Many kinds of anthropoid fossils have been found in deposits in Egypt, clearly showing that this superfamily of hominoids was established on the African continent at least 40 million years ago. They probably were present even earlier, since the first known hominoids already had diverged toward the separate gibbon family and the ancestors of pongids and hominids.

The recent discovery of fragments of the fossil species *Aegyptopithecus zeuxis* may indicate the presence of a progenitor stock that already was diverging toward the hominid line, thus providing additional evidence that gibbons were separating as a distinct group toward the end of the Oligocene epoch. Although *Aegyptopithecus* was as small as a gibbon, its jaws and teeth were more like those typical of apes than of gibbons. This interesting specimen is some 8 to 12 million years older than any other known type of generalized anthropoid fossil and provides evidence of a group that, though it was neither pongid nor hominid, may have continued to evolve toward these forms as the gibbon lineage became more divergent.

At the start of the Miocene epoch, about 25 million years ago, gibbons had diversified and probably had already spread to Asia and Europe as well as having extended their range in Africa. The particular form considered to be ancestral to pongids (great ape members of the family Pongidae) and hominids (members of the family of man, Hominidae) is the hominoid dryopithecine group. Dryopithecines have left a reasonably good fossil record from which it is known that they occurred in Africa, Asia, and Europe during Miocene times. Although relatively few different species have been recognized in the genus *Dryopithecus* in the fossil record, they all have rather generalized dentition and locomotor traits. The development of such specialized modes of locomotion as brachiation, knuckle-walking, and upright bipedalism, must have evolved later and perhaps from different dryopithecine lineages. The somewhat controversial species *Dryopithecus africanus* (formerly called *Proconsul*) from Kenyan deposits of Miocene age has been considered a likely ancestor of various later groups of hominoids. The species had small canine teeth and incisors and lacked the "simian shelf," a region of the lower jaw to which large and powerful muscles are attached in modern great apes. Also its arms were shorter than forelimbs in modern pongids, which may be indicative of a terrestrial existence in which the quadrupedal mode of locomotion was not exclusive; certainly the arrangement is not that found in brachiators, whose forelimbs are longer than their hindlimbs. The existence of such generalized features has proved tempting for speculations on patterns of evolutionary descent. Theories about an ancestral line that diverges to produce various life forms always are geared toward a generalized set of traits. It is easier to discern or explain different adaptive trends developing from a generalized set of traits, and it is by far the commonest pattern in evolutionary sequences. Yet despite the obviously generalized aspects of *Dryopithecus africanus,* there is little agreement concerning its particular role in hominoid evolution.

The Pliocene form *Ramapithecus,* found in deposits in India and east Africa, is considered a likely candidate for the ancestral type from which the hominids descended. *Ramapithecus* itself is believed to have been descended from a dryopithecine lineage, and its approximate age is 14 million years. Unfortunately the fossil remains consist only of some upper jaws and teeth, but these are reduced in size along lines of change that would be expected in archaic hominids. Nothing is known of the trends toward increasing cranial capacity or of locomotor modifications, but if *Ramapithecus* is accepted as diverging toward hominids and away from pongid specializations, our ancestry may be traced back to the beginning of the Pliocene epoch and to this particular species. If there was a divergence from a generalized hominoid to pongids and hominids at this time, this would imply an earlier form as the prior generalized ancestor, probably a Miocene form such as *Dryopithecus.*

Although the Pliocene epoch is crucial to our understanding of human ancestry, so few fossils have been found that it is not possible to delineate unequivocal sequences of descent and particular modifications that may have taken place during the 10-million-year interval. Some discoveries were interpreted quite differently when they first were found, compared with more recent ideas about them, and it seems likely that future discoveries will change some of the current notions about our ancestors. The "abominable coal man" (*Oreopithecus*) from Pliocene deposits in the Italian region of Tuscany once was considered to be ancestral to humans but now is accepted as an offshoot group, which developed a brachiating mode of locomotion. Since an entire skeleton has been preserved, the interpretations could be made more carefully than in cases where a tooth or bone fragment is the entire remaining specimen. Another Pliocene group was for a time assumed to be a hominid ancestor on the basis of reduced canine teeth, but additional fossil specimens clearly show that *Giganto- pithecus* species of northwestern India were an offshoot pongid line. In fact, the available evidence provides one of the best reconstructions for a lineage in ape evolution, as follows: over a span of 10 million years, the late Mio- cene species *Dryopithecus indicus* gave rise to the early Pliocene descendant *Dryopithecus giganteus,* which evolved by the mid-Pliocene to *Giganto- pithecus bilaspurensis* and thence to the mid-Pleistocene species *Giganto- pithecus blackii.* The forms then became extinct.

The most recent finds that may be hominids or immediate ancestors of hominids consist of two fossil fragments from Kenya. An elbow discovered at Kanapoi has been dated as about 4 million years old, and a lower jaw fragment with one molar still in place was found at Lothagam near Lake Rudolph. The Lothagam jaw is about 5.5 million years old and bears a close resemblance to australopithecine structures, which are definitely early hominid types. Except for these specimens, which may be interpreted in various ways, there are no other known fossil remains of Hominidae earlier than the Pleistocene epoch. We will consider the more recent history of the human family in the next chapter.

Some Suggested Readings

Clark, W. E. L., *History of the Primates.* 5th ed. Chicago, University of Chicago Press, 1965.

Colbert, E. H., *Evolution of the Vertebrates.* 2nd ed. New York, John Wiley, 1969.

Evans, L., "Ancestral secrets." *The Sciences* 12 (April, 1972), 16.

Pilbeam, D., "*Gigantopithecus* and the origin of Hominidae." *Nature* 225 (1970), 516.

Simons, E. L., "The early relatives of man." *Scientific American* 211 (July, 1964), 50.

Simons, E. L., "The earliest apes." *Scientific American* 217 (December, 1967), 28.

Simons, E. L., and P. C. Ettel, "*Gigantopithecus.*" *Scientific American* 222 (January, 1970), 76.

Tattersall, I., "Of lemurs and men." *Natural History* 81 (March, 1972), 32.

Tuttle, R. H., "Knuckle-walking and the problem of human origins." *Science* 166, 953.

Wessells, N. K. (ed.), *Vertebrate Adaptations: Readings from the Scientific American.* San Francisco, Freeman, 1968.

PART IV

HUMAN EVOLUTION

CHAPTER 13
ANCIENT
AND
MODERN HOMINIDS

Although the significant discoveries at Kanapoi and Lothagam may indicate the existence of members of the human family toward the end of the Pliocene in Africa, the information provided to date has been inadequate for us to be sure that was indeed the case. If these fossil fragments do represent ancient australopithecines, then we may date the first members of our family as long ago as 5.5 million years. Aside from these two specimens, the first fossils to be identified unquestionably as hominids are about 2 million years old, and they resided in southern Africa during the Pleistocene Epoch. The first report was made in 1924 by Raymond Dart, who described the fossil remains of the skull of a child, who was small and gracile and had rather small teeth. This specimen, which was found at Taung in South Africa, is representative of the species *Australopithecus africanus,* a member of the subfamily Australopithecinae of the family Hominidae. More than 300 specimens have been found since 1924, at eight different locations in southern and eastern Africa; a few fossils have been found also in Indonesia. Since whole skeletons have been found in some places, it is possible to make more accurate interpretations of the group and its distinctive features.

Two other species of *Australopithecus* have been recognized by some authorities, *A. boisei* and *A. robustus,* both being relatively large-toothed types of substantially greater bulk than *A. africanus.* There is some controversy concerning the correct affinity (hence, nomenclature) for the various australopithecine fossils, and different synonyms are widely used by different investigators. John Robinson recognizes *Australopithecus africanus* and *Paranthropus robustus,* as do others. According to Louis Leakey, one of the principal authorities in this field of study, the australopithecines were exclusively African in distribution. But other paleontologists, such as Elwyn Simons at Yale University, insist that members of this subfamily lived in Asia as well as Africa, since Indonesian fossils have been found.

The Earliest
Hominids

Why do we consider the australopithecines to be hominids and not pongids or even more generalized hominoids? Hominoids comprise a large category of anthropoid species that are quite distinct from the monkeys of the Old and New Worlds, these being the alternative anthropoid types. With sufficient information we can narrow down the identification of a particular specimen or species to one of the three hominoid groups: gibbons (hylobatids), great apes (pongids), and human or prehuman types (hominids). The gibbons diverged early to various brachiating forms and are different in numerous ways from pongids and hominids. The principal distinction, among many others, that serves to identify a specimen as hominid rather than pongid is the development of upright posture and bipedal locomotion. This trait occurred early in hominid evolution and serves as a unique index to identify members of our own family as different from species of great apes and their extinct relatives. The excellent fossil remains of the australopithecines clearly indicate that the group developed upright bipedalism, unlike the quadrupedal or brachiating pongids. On this basis alone we may be able to classify an unknown form or fragment as hominid, if the fossil deposit is of an appropriate age relative to the presumed evolutionary progression.

There are specific anatomical traits that clearly indicate the feature of upright bipedalism in fossil hominids (see Fig. 12.12). The ilium of the pelvis is shortened and pushed back in hominids as compared with pongids. This shape and positioning provide improved support for the abdominal organs of an upright creature. New muscle and ligament attachments also are required for such a pelvis to be successful. The upright species can extend the femur completely, thus straightening the knee. The bent-knee walk of pongids reflects their inability to accomplish these same actions, because muscle and ligament attachments are different. The gluteus maximus muscle extends from the sacral part of the human pelvis to the upper femur and permits the complete extension of the femur; this finishes the swing and the knee straightens. The gluteus maximus muscle is quite small in great apes, and rather than extending the femur it functions to abduct the thigh. Thus this muscle is an extensor in hominids but serves as an abductor in pongids. Humans whose gluteus maximus muscle has become paralyzed are obliged to walk with the bent-knee gait of the higher anthropoids. This muscle contributes to the more conspicuous buttocks of humans than one finds for ape species.

The hominid foot became modified as a support structure and it is substantially different from the foot of the quadrupedal pongids (Fig. 13.1). The pivotal position of the great toe leads to a particular wearing pattern of the bone; therefore, a single fossil great-toe bone can indicate bipedalism

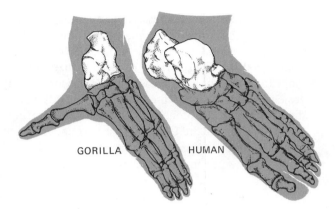

Figure 13.1 Comparison of bones of the foot in pongids and humans.

GORILLA   HUMAN

and the striding walk of its former owner. In addition to this, the great toe no longer is opposable in hominids as it is in pongids and other primates. Other digits of the foot also have become reduced in size and have essentially lost all or most of the grasping capacity that still characterizes other primates.

The relative lengths of forelimbs and hindlimbs differ in hominids and pongids. The modern orangutan is an arboreal brachiating form with the very long forelimbs characteristic of species that utilize this mode of locomotion. Bipedal humans have relatively shorter forelimbs than chimpanzees and gorillas, whose forelimbs, in turn, are proportionately shorter than in the brachiating species. The chimpanzees and gorillas are "knuckle-walkers" whereas the orang is a "fist-walker." These differences presumably arose later in pongoid evolution and appear to be related to the increasingly longer intervals spent on the ground during the daily activity cycle of chimpanzees and gorillas. The arboreal orangutan rarely leaves the trees for any reason, and its muscle and ligament arrangements represent adaptations to habitat and to locomotion in that habitat.

The skull is positioned above the backbone in hominids; hence the more central position of the foramen magnum (see Fig. 12.11) clearly indicates continuous upright posture. The foramen magnum in pongids is located more toward the rear of the skull, indicating a different relative position of the head over the backbone that is typical of these quadrupeds. Of course the great apes can assume an upright posture and walk on two legs as well as on four. But the principal habit is quadrupedal in pongids, whereas hominids consistently maintain upright posture and walk on two legs all the time.

A variety of other characteristics provide equivalent indication of differences between hominids and pongids so that quite fragmentary remains may provide clear evidence of fossil identity even when the locomotor apparatus is unknown or absent. The shape of the palate is rounder in hominids, and the jaws tend to be less massive than in ape lineages. In pongids the canine teeth are quite large and pointed, and there is a gap between the incisor teeth in front and the canine teeth immediately behind. In hominids the canine teeth are considerably reduced in size, often resemble the incisors in shape, and are not separated from the incisors by any large space. The face projects forward under the brows in pongids more than in hominids; the shape of the palate alone may provide sufficient information to reconstruct the dimensions of the face, because a rounder palate leads to a flatter face while the longer pongid palate contributes to its more muzzlelike face. The cranium is larger in hominids than in pongids, reflecting larger brain size and a more substantial cerebrum accommodated within a higher cranial vault. In modern humans the cerebrum essentially overlies the rest of the brain. The pongid thumb is more of a "hook" and represents an adaptation that permits rapid release of the hand from whatever it holds onto, whereas the longer thumb in humans has grasping as well as "hook" capabilities. The human thumb is more opposable and permits a wider spectrum of activities for the hands than can be achieved by pongids.

Although they possessed upright posture and the ability to walk on two legs, australopithecines are distinguished from more advanced hominids by the lack of certain significant characteristics. The principal difference is the relatively small size of the australopithecine cranium, which indicates that its brain was not much larger than that of modern pongids (Fig. 13.2). Indeed it was because of these fossils that we first came to realize that the evolution of hominids involved changes in posture and locomotion long before there was an increase in brain size (intelligence). The earlier notion that human ancestors were "bright apes that came down out of the trees" had to be abandoned, and we now recognize quite clearly that locomotor modifications preceded brain changes during hominid evolution. The skull was very thick and the brow ridges were projecting in the australopithecines. Although the teeth were more like those of hominids than of pongids in shape and spacing, they were considerably larger than those of more advanced hominids. Along with larger teeth, the jaws were massive compared with later hominids, but they were not as large as the jaws of pongid species.

There is no compelling evidence that australopithecines were capable of *making* tools; certainly the small size of the brain is not. They did, however, probably use tools. Tools made of horn and bone have been found in fossil deposits, but they may have been used without any more modification than is achieved by some animals that use sticks or twigs or stones to accomplish some immediate task. There is no evidence of a tradition or progressive improvement in these tools. Hominines are now defined as tool-makers, or, more particularly, as species that use tools to make other tools. This more specific definition was established as it became obvious that various animals were fully capable of utilizing some natural objects as tools. Sometimes the object is unchanged and sometimes it is refashioned in some

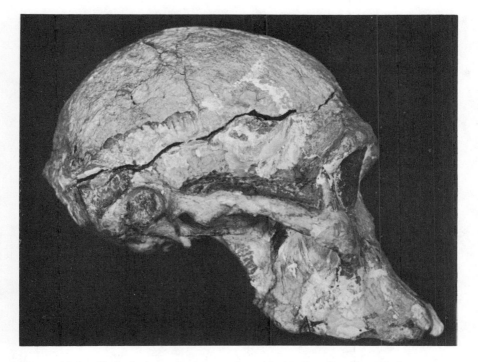

Figure 13.2 Skull of *Australopithecus* found in 1947 at Sterkfontein in South Africa. (Photograph courtesy American Museum of Natural History, New York.)

way to make it more effective for the particular use. An especially pertinent observation was reported by Jane van Lawick-Goodall for the activities of chimpanzees in natural populations in central Africa. She observed repeatedly that chimpanzees would select branches, strip off the leaves, perhaps change the length of the twig, and then probe into termite nests and eat the termites that came out sticking to the tool. This led to a redefinition (a refinement of the original definition, according to one view) for tool-*making* versus tool-*using*. Many other species use tools. The woodpecker finch of the Galapagos Islands selects cactus needles or twigs of a proper length (avoiding other lengths) and uses these to probe into places in trees where insects occur. The sea otter lies on its back in the water and uses a stone to smash open hard-shelled food organisms, which it balances on its belly during the operation. But no tool is used by such species to make the final tool, whereas ancient man used stones to fashion other stones for special purposes, or a variety of tools to create others for particular functions. Since the current belief is that australopithecines used existing objects with little or no modification, they do not qualify as toolmakers. That designation is reserved for hominine members of the Hominidae.

One other feature that makes the australopithecine different from the more advanced hominine species of hominids is his apparent lack of culture. Culture has many meanings, but basically it refers to the transmission of learned knowledge from generation to generation. More particularly, we expect an accumulation of this knowledge to take place so that each generation does not have to discover or invent anew what the previous generations possessed as skills and other forms of learning and tradition. Inasmuch as we cannot detect a particular style of tool design in australopithecine deposits, it seems unlikely that these hominids possessed a culture equivalent to those associated with later species. The random nature of the tool characteristics suggests that there was no pattern of accumulation and transmission of accumulated knowledge, nor improvement in the body of knowledge. Thus australopithecines may resemble those animal species that acquire a behavioral trait by mimicking, such as the observed local tradition of washing potatoes before eating them which has become established in some colonies of macaques.

From this brief description it is quite clear that hominid fossils may be identified from appropriate remains, no matter how fragmentary. For more certain identities it is desirable to have many specimens and many structural and artifactual materials to reconstruct the appropriate creature and its life style. The jaw fragment with a premolar in place is all the evidence now available for the putative 5.5-million-year-old australopithecine found at Lothagam in Kenya. This much information is inadequate for the moment to provide a preliminary niche in the lineage of the hominids. Much more substantial evidence will be required to locate those forms that existed during the 10 to 12 million years that intervened between *Ramapithecus* and *Australopithecus africanus*.

The Hominine Species

Except for specimens of more recent populations and occasional fragmentary remains from Asian sites, little was known for a long time about the sequence of forms that comprised the hominine subfamily and led to the ultimate appearance of modern humans. After the discovery of australopithecine fossils in Africa in 1924 and subsequent finds in that continent, the focus of search shifted there from Asia. The important point was that the African finds were much earlier so that it seemed a more probable locale for the origins of the human family. Significant fossil specimens continued to be found, but extremely important discoveries especially pertinent to our current notions about the "missing links" between australopithecines and hominines, were made by Louis and Mary Leakey and their colleagues in the Olduvai Gorge in the Tanzanian region of east Africa. Controversies developed, as usual, and the clarity of the evolutionary picture is not as sharp as it could be, but there are some compelling lines of evidence in support of the interpretations made by the Leakeys.

Leakey considered the newly named species *Homo habilis* to be the Oldawan toolmaker whose remains frequently occur together with australopithecines. It is not altogether certain yet, but the available evidence is that the tools are found only when *H. habilis* remains are present, whether or not there are also australopithecine artifacts. Deposits that have only australopithecine materials never contain the Oldawan "pebble" tools made of stone. On the basis of such correlations, Leakey has suggested that the habiline species is a member of the genus *Homo,* toolmakers, whereas *Australopithecus* had not yet reached the hominine achievement level. This first of the known hominines was a pigmy-sized species that had a larger cranial capacity than either the australopithecines or pongids. The species had reduced and narrow teeth (Fig. 13.3) and limb features indicative of upright bipedal forms. The tools found with habiline fossils appear to have required the use of other stones to be manufactured, and the occurrence of a sequence of improved patterns over many generations provides further evidence of the cultural basis to the toolmaking tradition. The particular significance of dental and cranial traits is that each shows intermediate values between australopithecine measurements and those for *Homo erectus,* a definite hominine species of a later time. This quantitative relationship was predicted from the specimens known before the discovery of *Homo habilis,* so that the fit seemed real and not merely fortuitous. The mean cranial volume of australopithecines was 500 cubic centimeters (cm³) and about 1000 cm³ for *Homo erectus;* thus an average of 680 cm³ for the habiline fossils indeed seemed to be a reasonable intermediate size for the evolutionary sequence.

Figure 13.3 Lower jaw and teeth of the hominine species *Homo habilis* found in the Olduvai Gorge region in Tanzania. (Photograph courtesy of P. V. Tobias. Reprinted with permission of the author and the publisher from "Early Man in East Africa" by P. V. Tobias, *Science*, Vol. 149, Fig. 8, pp. 22–33, 2 July 1965. Copyright © 1965 by the American Association for the Advancement of Science.)

The evidence is fragmentary, but the available information has been put together to construct a working hypothesis that early Pleistocene hominids of different species coexisted and evolved concurrently. Some were of a more advanced type than others, and some were adapted to different living zones or life styles. Since different fossil forms have been found in the same locales, it seems possible that the ancestral form was of Pliocene age and that it diverged to produce the two or more australopithecines found in Pleistocene deposits. Perhaps the Kanapoi and Lothagam fossils will provide more information about the Pliocene ancestors of the Pleistocene species.

One sequence proposed by John Robinson and others is that the more robust australopithecine (*Paranthropus robustus* or *Australopithecus robustusboisei*) was present earlier and that its habits and living pattern were more apelike. Individuals weighed about 150 to 200 pounds and had a mixture of cranial, facial, and dental features indicating a herbivorous diet and a terrestrial existence similar to modern chimpanzees or gorillas, that is, lush tropical habitats that permitted much time on the ground but nighttime retirement to the trees. The *robustus* type had a somewhat divergent great toe and impressive teeth for grinding and crushing foods, aided by a more massive masticatory apparatus (Fig. 13.4). Despite the fact that the great toe was somewhat divergent, the *robustus* species had other, upright bipedal traits, though they were not as well developed as in the gracile *Australopithecus africanus*. Robinson has suggested that the robust line continued for perhaps 2 million years in a relatively stable state, but another line diverged to produce a smaller form with different habits and appear-

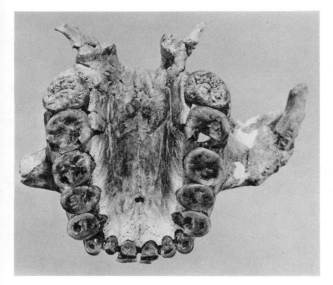

Figure 13.4 Teeth and upper palate of the robust type of australopithecine, from the Olduvai Gorge region in Tanzania. (Photograph courtesy of P. V. Tobias. Reprinted with permission of the author and publisher from "Early Man in East Africa" by P. V. Tobias, *Science*, Vol. 149, Fig. 3, pp. 22–33, 2 July 1965. Copyright © 1965 by the American Association for the Advancement of Science.)

ance. The gracile form was established by the early Pleistocene, according to this idea.

Australopithecus africanus weighed about 50 to 60 pounds and may have been a hunting-and-gathering species, as evidenced by the varied teeth for shearing, piercing, and gripping as well as grinding and crushing, and by horn and bone artifacts found with bones of small animals (lizards, suckling animals, and others). Such presumptive remains of food materials are similar to the animal food supplements used by hunting-and-gathering human societies today and would have been available to the ancient residents of the open savannahs. Although the absolute cranial volume was about the same for the robust and the gracile species, there was a larger brain proportionate to body size for the *A. africanus* individuals, which could indicate its greater intelligence. The greater intelligence is not certain, however, because the specific development of the brain is more important than absolute or relative size even in the same species; for example, modern humans with a brain volume of 1000 cm³ are not necessarily less intelligent than others whose brain volume may be 1800 cm³. According to this hypothesis, the gracile australopithecine type is the more probable candidate for ancestry of the first hominines, and the robust species is considered to have become extinct and left no issue in later times.

Alternative ideas about these hominid ancestors generally agree that *A. africanus* is in the direct lineage to the hominines, but disagree on other aspects of the evolutionary sequence. Phillip Tobias, a colleague of the Leakeys, has proposed that *A. africanus* was the progenitor species for the Pleistocene australopithecines and hominines. The *A. africanus* populations thus would have diverged to produce, in one lineage, the large-toothed, robust, somewhat apelike species (or two similar large-toothed species, *A. robustus* and *A. boisei*) which ultimately became extinct, while it would have retained a second, gracile lineage that evolved to the level of the first hominine species with which it had much in common. The size, dentition, cranial features, habitat preference, and other characteristics of *Homo habilis* represent a reasonable continuum with *A. africanus*. The significant increase in mean cranial capacity and the clear evidence of a cultural toolmaking tradition provide the basic criteria for the separation of australopithecines and hominines. With a greater sampling of *H. habilis* fossils it may become even more apparent that the dividing line between the ancestral and derived forms was quite gradual and subtle. But the beginnings of a cultural capacity provide such substantial dimensions for evolutionary progress and development that we would maintain the classification dichotomy that distinguishes the two hominid subfamilies.

Although we are uncertain of the cause of extinction of these ancient groups, intergroup competition for common resources has been cited frequently as the factor responsible for the replacement of species by more advanced forms that have similar habits and preferences and occupy a common adaptive living zone. The relationship between Asian and African fossil hominids is not very clear. Perhaps the species arose in one area and migrated subsequently to occupy an extended range. After becoming established in different regions, such populations might have evolved along similar pathways in response to similar selection pressures acting on a common genetic background. Such instances of parallel evolution are entirely possible, but there is little evidence to permit any decision on the processes that contributed to the development of scattered populations along very similar evolutionary lines.

Although *Homo habilis* is a controversial species, there is general agreement on the identification of *Homo erectus* as an unequivocally hominine form. The first discovery of this species was made in Java in 1891 by a Dutch anthropologist named Dubois, the fossil specimen consisting of a skull and thighbone about 700,000 years old. This so-called Java man was given the formal name of *Pithecanthropus erectus,* but this was changed subsequently to *Homo erectus* in accord with standard rules of nomenclature and the bases for classification of Hominidae. The fossil Peking man, formerly called *Sinanthropus pekinensis,* represents a more recent and advanced stage of *Homo erectus* evolution. Despite its larger cranium and other improvements over the more ancient Java man, the two types are sufficiently similar to be considered members of the same species. These specimens were the only ones known until about 25 years ago, so until that time human origins were considered to be more restricted geographically than we know to be the case today. Swartkrans man was discovered at South African sites in 1949, and the type was named *Telanthropus,* although it now is classified as *Homo erectus.* A specimen of *H. erectus* about 500,000 years old was uncovered in the Olduvai Gorge in east Africa in 1961. Lantian man, found in China in 1963–1964, appears to be an older form of *H. erectus*; that is, it is more like Java man than like its neighbor Peking man. In 1965 there was a report of an advanced form of *H. erectus* from Hungary, but its affinities are less certain at present. Many specimens of hominine fossils, which were stored in a large collection at the University of Peking, were lost during the Chinese revolution in 1949. Whether these priceless remnants of our history will ever be found and made available again for study is an unanswered question today.

Homo erectus apparently lived at least in Asia and Africa and probably in Europe for a period of about 200,000 years (Fig. 13.5) but disappeared some time in the early part of the mid-Pleistocene, some half million years ago. In its physical features *H. erectus* had a skeleton much like ours, but the brain was smaller, the skull was flatter and thicker, brow ridges protruded in front and were angular in the rear, the teeth were larger and more primitive, and its stature was shorter than in modern humans, on the average (Fig. 13.6). There was a definite culture, as seen from artifacts found with the hominid fossils, including evidence for the use of fire. There has been no evidence to indicate that fire was used by the australopithecines or

Figure 13.5   Distribution of *Homo erectus*, based on known fossil fcrms discovered at these sites.

by *Homo habilis*. A number of the skulls found in the cave sites associated with *H. erectus* apparently had been bashed open. It is quite possible that this was a tradition similar to practices of some present-day human groups, which remove the brains of their victims for different superstitious reasons. More importantly, the presence of bones of many large animals in these sites provides the first evidence that big-game hunting was a way of life in prehuman species. All theories about a predominant hunting life style earlier than this have been pure conjecture unsupported by evidence from the fossil record. The hunting-and-gathering economy is quite different from one in which the major activity is cooperative hunting for large animals. We will discuss the significance of these distinctions in Chapter 14 when we consider social traits in relation to the development of sex roles in prehuman and human societies.

The affinities of a number of fossil forms are poorly understood. Solo man from Java and Rhodesian man from Africa have some primitive features but are considered unlikely to be specimens of *Homo erectus* because they existed during the late Pleistocene, hundreds of thousands of years after the disappearance of this species from other parts of the world. It is more likely that these two specimens represent Neanderthals of a later time than that they are examples of relict populations of *H. erectus* persisting almost a half million years after the rest of this widespread species became extinct. A lower jaw with teeth in place, designated as Heidelberg man, presents difficulties in classification because of its rather small teeth but relatively massive jawbone. Perhaps it represents a type that was transitional between *H. erectus* and *H. sapiens,* but we do not know. Other fossil specimens, found at Ternifine in Algeria, also are about 500,000 years old and have the same massive lower jaw features as the Heidelberg fossil and the Olduvai specimen; all are of the same age and all are considered to be *Homo erectus* or to have close affinity to this species. The observed variations in these fossils may reflect nothing more than the range of variability one would expect in separate populations of a widespread species that evolved over an interval of 200,000 to 300,000 years. Since their age and geographical distribution coincide with the occurrence of known *Homo erectus,* it seems most likely that they belonged to slightly different individuals of the same species.

The problem of Piltdown man was solved some time ago, but it presented great difficulties when it was discovered in the nineteenth century, and for some years afterward. In earlier days before we had substantial evi-

Figure 13.6 A reconstructed skull of *Homo erectus* from China, formerly called Peking man. (Photograph courtesy American Museum of Natural History, New York.)

dence that locomotor modifications preceded brain increase in hominid evolutionary history, it was considered that the brain increased in size and that upright bipedalism evolved afterward. The Piltdown skull was found in two separate parts, near each other; it consisted of a decidedly simian jaw but a cranium that was very much like a modern human's. In addition, the simian jaw was a puzzling find in a deposit of Pleistocene age in England. As more and more evidence was accumulated in the delineation of hominine history, the Piltdown fossil became a greater embarrassment. After its discoverer had died, dating analysis was performed and it was shown very clearly that the upper and lower parts of the skull were completely unrelated and the presumed fossil was a fraud. The reasons for the hoax are unknown. Although the fossil has no scientific credibility, it does provide an example of how objective methods can be used to validate or discredit ancient materials. We can thus be more confident of fossil specimens that have been subjected to stringent tests, although subjective interpretations will continue to becloud the relationships of various particular samples.

Neanderthal hominines first were discovered about 100 years ago in a valley of that name near Düsseldorf in Germany. Many additional specimens have been found since then in various parts of Europe, Asia, and northern Africa. Fossil Rhodesian and Solo specimens, as mentioned above, may have been Neanderthals with some differences relative to most of the European materials, which form the bulk of Neanderthaloid collections. There may have been an earlier Neanderthal type with more generalized features, but most of our information has come from studies of the later forms, which possessed some very distinctive traits. Although the mean cranial capacity was slightly higher than in modern *Homo sapiens* (1400 versus 1300 cm³), the skull had massive features such as protruding brow ridges and points of attachment for powerful shoulder muscles. The forehead was retreating above the massive brow ridges, the jaws were large and projecting, and there was no chin eminence. Some limb bones were unusually thick and showed curved shafts, which may have been due to some disease such as arthritis in some members of these populations. Other known Neanderthal populations showed skull features much like those of modern human beings and limb bones that were distinctly *sapiens*-like. It would seem reasonable to interpret these data as showing that an earlier Neanderthal type existed, during the final interglacial and the early part of the last glaciation of the Pleistocene epoch, and varied more or less as would be expected for small, separated populations of a sexually reproducing species. This forerunner type may have diverged to produce an offshoot "extreme" line along with others that were still rather generalized in all or most traits. The "extreme" Neanderthal populations of central Europe were quite homogeneous, and would thus be more likely to represent a divergent group with characteristics outside the range of, and distinct from, modern humans or of the generalized Neanderthal groups. In any case, Neanderthals disappeared abruptly from Europe during the last glaciation, about 50,000 years ago, and were replaced by *Homo sapiens* groups. These populations of modern humans may have evolved from the generalized type that also produced the "extreme" Neanderthals, or there may have been a common ancestral lineage, which produced Neanderthal and sapiens types separately. Indeed there is some evidence from analysis of the earlier Pleistocene populations that variability ranged from individuals with more Neanderthallike traits to others with more *sapiens*-like traits in various communities. Divergence to produce the two separate forms therefore may have occurred in those groups identified as belonging to early Mousterian cultures. The "extreme" Neanderthals of central Europe are considered to represent the late Mousterian stone tool culture, which was highly developed and occurred later in the last glaciation episode, coincident with the final appearances of these now-extinct populations.

Homo sapiens, modern humans, occurred at least 40,000 years ago and probably earlier than that. Evidence from radiocarbon datings, faunal associations in the fossil deposits, tool traditions, and other artifacts, demonstrate that humans populated regions of Europe, Asia, and Africa for tens of thousands of years. Artifacts of the Aurignacian culture have been dated as 25,000 to 30,000 years old. This cultural tradition is associated with *Homo sapiens*; it essentially replaced the Neanderthal Mousterian culture at the same time that changes occurred in the physical forms of the two groups. These Paleolithic populations were replaced in turn by members of the more recent groups associated with the Magdalenian culture, such as the Cro-Magnon type, which is indistinguishable from modern Europeans. The Magdalenian culture counts among its remains the exquisite paintings in caves of southwestern Europe, such as the Lascaux Caves in France. Some authorities consider it possible that the Basques of southwestern Europe and the Berbers of northwestern Africa may represent persistent Cro-Magnon groups, or immediate descendants of these peoples of the Magdalenian paleolithic culture who existed in southwestern Europe about 15,000 years ago.

Homo sapiens populations are characterized by more efficient upright posture and bipedal locomotion, a prominent chin, brows that hardly protrude, reduced and shortened jaws and teeth, a rounded cranium, and other traits of general occurrence. The species is worldwide in distribution and has been so ever since the beginnings of recorded history, except for recent transients in Antarctica. From the fossil record we know that human groups migrated into Australia about 16,000 years ago or earlier, and into North America and then to South America anywhere from 25,000 to 50,000 years ago. As newer archeological discoveries are made, especially in the western United States and in South America, the absolute date is pushed further back. The mongoloid groups, which entered the Americas from Asia, via the Bering Strait, are different from other members of the ancestral lineages who remained in Asia, each group having undergone different adaptive changes. Aboriginal populations entered Australia from Asia, and we believe that no other substantial migrations took place there until 1788 when the British landed at the site of present-day Sydney on the east coast of the continent.

From what we can detect in prehistoric sites, the change from a primarily hunting-and-gathering life style to an urbanized way of life may have begun some 9000 years ago or earlier, with the introduction of agriculture. Many societies continued the nomadic or stable nonagricultural existence, as we know quite directly for various peoples in Australia, Asia, Africa, South America, and North America, today and in the recent past. Many of the North American Indian groups were nomadic until their lands were taken from them during the settling of the West. Primitive tribes exist in various, usually inaccessible, regions of the other continents, including the Tasaday people recently discovered to be living in caves in the forested regions of the Philippine island of Mindanao.

The two principal areas of origin of crop plants and domesticated animals are considered to be the Near East and the central New World. Such staple foods as wheat in the Near East, the sweet potato in Africa, and the date palm in India, were developed along with domesticated dogs, horses, cows, pigs, sheep, goats, chickens, camels, and donkeys in the Old World. The dog may have been the first animal to be domesticated, as early as 7000 B.C., but we know that the camel was absent from the art work of the early Egyptian dynasties and that it came under domestication at a later time. In addition to the introduction of corn and potatoes in the agriculture of Central and South America, the turkey also was used as a food source, and the llama of South America was a beast of burden and source of clothing equivalent to the various animals of the Old World societies. The implications of these cultural developments for the evolution of human societies will be explored in the next chapter. But it is quite clear that the development of the Neolithic traditions of the Near East led to larger population groups in more settled circumstances. The raising of grain and simple animal husbandry, along with the making of pottery and cloth, gave way to more sophisticated accomplishments in earliest historical times. The ancient Egyptians and Mesopotamians were still Neolithic, but about 5000 or 6000 years ago copper and then bronze were in use for making tools, weapons, and various sorts of vessels. This was followed by the initiation of the Iron Age in the Near East, about 1400 B.C., when humans acquired the skill and knowledge to extract iron from ores. The invention of the wheel and such means of transportation as the dugout canoe were Neolithic; sophisticated modifications for some of these creations did not appear until very recent historic times.

From the discussion so far, we can summarize the significant biological features of humans as distinct from the procession of hominid and hominine ancestral and related forms. We share with other hominids the capacity for continued upright posture and bipedal locomotion, although improvements have been incorporated since these traits first appeared in the earliest Pleistocene groups. The development of culture distinguishes hominines from australopithecines, and can be traced back more than 2 million years in our history. While upright bipedalism reflects predominantly anatomical and physiological genetic modifications, cultural characteristics depend not only on physical dexterity but also on the increased intelligence of the larger-brained hominines. Increased intelligence is genetically determined, principally, but there is substantial nongenetic input to the total exploitation of the cultural potential. The ability to learn and to accumulate learning over the generations provides an evolutionary shortcut, which permits a dazzling speed of improvement and change totally unknown for traits based exclusively on genetic factors. The acquisition of culture represents an evolutionary innovation of unparalleled potential and certainly specifies a uniqueness for humans in the biological world.

Another cultural trait, which depends in part on genetically specified muscle, bone, and brain tissues organized in special ways, and in part on learning during one's lifetime, is the human ability to communicate symbolically by language. Other animals can use a variety of signals, some of them vocal signals, to communicate information, but they seem incapable of transmitting abstractions and complex ideas. Recent reports of Washoe and other chimpanzees, which learn to "speak" using sign language, may lead us to redefine humans as a uniquely linguistic species, just as we tightened our claim on toolmaking when we discovered that other animals also used tools. For the moment, however, there is general agreement that the human species is the only one that communicates linguistically in *all* its societies. Each of these traits depends on genetically specified biological systems, but the cultural properties of toolmaking and language provide a dimension of evolutionary improvement that permits the human species a dizzying spectrum of potential advantages that cannot possibly be realized by nonhuman animals.

The Human Races

To discuss the races of the human species we first ought to review the concepts of populations and speciation, since humans are not unique in having a compartmented species complex. The conventional criterion of a species is that gene flow can occur freely within and between its populations, but interbreeding either cannot take place or rarely occurs between populations of different species because of the level of genetic differentiation that prevails. The underlying premise is that the genetic differences prevent mixing of gene pools so that these remain distinct from one another in cases of separate evolutionary groups. Because there is a gene pool for each species, particular kinds of genes and particular frequencies of alleles and genotypes will be present and thus distinguish different species. The genetic differentiation that leads to eventual reproductive isolation of species becomes established over long periods of time, perhaps tens of thousand of years for higher life forms with longer generation times. Different populations of the same species also may show different allele and genotype frequencies, but the total gene pool is accessible to all the component populations theoretically if not in actuality. The greater the diversity in the gene pool, the greater the evolutionary potential is considered to be for the species. If genotypes of adaptive value arise in one population, they may be incorporated into other populations relatively quickly, depending on the prevailing selection pressures, because recombination and gene flow can occur. As we discussed in Chapter 8, the compartmentalized population structure is considered to be the one with the highest probability for success over the long periods of evolutionary time.

There are no sharp definitions or criteria for subdivisions of a species, nor are there terms that are applied universally to such species segments. Among many plant and animal species that have been studied, geographically separated population groups have been called *subspecies*. The subspecies of a species are fully capable of interbreeding, but such intermingling ordinarily is restricted to the boundaries where the separate populations overlap. The production of even a few fertile, viable hybrids provides evidence that the restricted gene flow is a function of geography and not of potential for gene exchange. The term "variety" has been used to designate recognizably different strains of plants, and there is a range of genetic difference among such forms. Similarly, we refer to different "breeds" of horses or dogs or cattle, by which we mean genetic varieties within a single species with unrestricted potential for gene exchange. The term "race" is fully equivalent to "variety" or "breed" or "subspecies," depending on the usage in vogue and the geography of the group, although it has come to be used more for the human species than for others. All these designations refer to Mendelian populations, which differ from one another as a function of the relative frequencies of alleles and genotypes in the common gene pool of the species complex. But evolutionary change is continuing and dynamic, and totally random gene exchange usually is not realized; hence frequency variations remain to a greater or lesser degree in the compartments of the species. Each species is a composite of Mendelian populations, each of which arose during the history of the group as a consequence of the varying mutational events and selection pressures that prevailed.

Races therefore may arise according to the usual evolutionary processes by which the diversity that is generated becomes fixed in the direction of adaptation under the guidance of natural selection. The development of distinguishable Mendelian populations, or races, requires some interval of effective isolation or some isolating factor that prevents or restricts the interbreeding potential between groups. This would permit genetic differentiation of each group, because little or no opportunity for randomized interbreeding would occur to reconstitute a single accessible gene pool. Given sufficient time in isolation, races may continue to evolve toward reproductively isolated species in all life forms, including humans.

There is no agreement among anthropologists about the number of different human races. Some recognize as few as 3 races, while others may describe 30 or 50 or 200. The basis is essentially arbitrary and depends on undefined levels of difference, which are designated for the Mendelian populations in question. The Caucasoid ("white"), Negroid ("black"), and Mongoloid ("yellow") peoples are the minimal 3 races recognized by some authorities, whereas others consider that each of these 3 groups consists of racial conglomerates, which can be separated still further. It is possible to recognize any number of subgroups within each of the major groups, depending on the point of view of the observer as much as on the actual differences in these compartments. Caucasoid populations may be subdivided to describe various European types (northern, central, eastern, Mediterranean, Iberian, etc.) as well as groups from North Africa and people of western Asia, such as those in Turkish, Iranian, and Arab regions. Negroids include various groups of sub-Saharan Africa, such as the Congolese and other west Africans, the Bantu of the southern portion of the continent, the peoples of the Sudan, Ethiopia, Kenya, and other parts of east Africa, the forest pigmy of equatorial Africa, the Hottentots of southern Africa, and others. Populations from the far western Pacific may also be included, such as the inhabitants of New Guinea and the islands of Micronesia. North American blacks have a primarily African heritage but with a significant Caucasoid component (about 20 percent) in the gene pool, whereas South American Negroes, the Cape Colored of South Africa, and Negroid populations in other regions have different degrees of genetic admixture, and genes from different sources have in each case been inserted into the predominantly African genetic background. Mongoloids often are subdivided at least to distinguish the indigenous Indians of the Western Hemisphere. But they also may be divided further into Northern Chinese; the peoples of Siberia, Manchuria, Korea, and Japan (the "classic Mongoloids"); the inhabitants of southeast Asian regions; the Tibetans; and the Eskimo of North America.

Other racial groups that have been designated include the Capoids or Bushmen of the Kalahari Desert of southwest Africa, a diminishing descendant group of an ancient population; and the Australoids, which comprise the aboriginal populations of Australia, parts of New Guinea and the Malay peninsula, and various regions of the far western Pacific. Less frequently, but often enough to be cited, racial designations have been applied to various Pacific Ocean populations such as Melanesians, Micronesians, Polynesians, and Negritos. Two groups have been recognized in populations of the Indian subcontinent, the Aryan inhabitants of Pakistan and much of India as one type, and the darker Dravidian people of south India as another. The people of Ceylon are descendants of earlier migrations from India, but the first migrants came from northern India and a more recent group is descended from south India groups. The Ainu of northern Japan may be considered Caucasoid in descent and quite distinct from the more recent Mongoloids of the country. Other populations also have been described and differentiated on the basis of criteria similar to those used to characterize major and minor groups such as the ones mentioned above.

The numbers of genes that determine such morphological traits as skin color, hair color and texture, stature, fingerprint patterns, and some others, are not known with certainty. But several or many genes must be involved with each trait, according to the genetic and statistical studies that have been reported. While there usually is a prevailing trend for a particular characteristic, there is considerable variation within populations and races for all the genetic traits that have been analyzed. Among members of the major racial groups there is *continuous variation* in skin color from lighter to darker, in stature from shorter to taller, and in other features as well. Furthermore, it is well known that nongenetic influences may modify the expression of these genetically determined human characteristics. More useful as indicators of gene flow are traits that show *discontinuous variation,* that is, discrete classes without intermediate forms. Such characteristics may be functions of one or a few genes, or, in some cases, multiple allelic forms of a single gene may be involved. In any case, the expression of the trait is genetic and uninfluenced by usual modifying environmental factors.

Two examples of discontinuous traits, one multigenic and the other due to a few genes, are the fingerprint patterns and various blood group factors, respectively. Since these characteristics do not change during the life of the individual and are altogether free from modification by outside agencies, they serve as indicators of gene exchange between populations and of differing allele frequencies in gene pool compartments. Each phenotype is clearly distinguishable from its alternatives and each is the result of the action of a specific and different genotype. There are three patterns of fingerprints, which occur in all major racial groups but with different average frequencies (Fig. 13.7). The loops are preponderant in most Caucasoid and Negroid groups; Mongoloids show more whorls than loops, as do Australian aborigines and Melanesians. Bushman populations are unusual in having a frequency of arches as high as 10 percent, whereas most other racial groups either have no individuals with arches or considerably fewer than 10 percent have this fingerprint pattern (Table 13.1).

The genetics of the blood groups is relatively simple and well under-

loop   whorl   arch

Figure 13.7   Three basic types of fingerprint patterns in humans.

stood. Each combination is readily distinguishable; there is no change during the life of the individual; and no known factor influences the expression except the genotype. The A-B-O major blood group factors are distributed such that considerable variation exists between populations (Table 13.2). On the average it is the O-factor that is most frequent in humans and may be present in 100 percent of some American Indian populations. East Asian groups show a higher frequency of B-factor than occurs elsewhere, while Caucasoid and Negroid populations tend to contain more individuals with A-factor than with B-factor. Other blood group components include the Rh-factor and the Duffy factor (Table 13.3). The Rh-factor is more prevalent among Negroids than among Caucasoids and occurs in virtually every individual from American Indian and East Asian populations. Thus, Rh-positive individuals are less frequent among Caucasoids than in the other groups. The Duffy factor rarely is found in Negroids but is present in most East Asians, and varies considerably in populations of Caucasoids and American Indians. In conjunction with other evidence, combinations of blood group factors clearly provide indications of different allele frequencies among human populations, showing greater variation between races than within a race. Although the interpretations are trickier, such data may pro-

TABLE 13.1   RANGE (IN PERCENT) OF TYPES OF FINGERPRINTS
AMONG POPULATIONS*

Population Group	Loops	Whorls	Arches
Caucasian Europeans	63–76	20–42	0–9
Negroes	53–73	20–40	3–12
East Asians (Chinese, Japanese, and others)	43–56	44–54	1–5
American Indians	46–61	35–57	2–8
Australian aborigines	28–46	52–73	0–1
Bushmen (Capoids)	66–68	15–21	13–16

*Adapted from *The Living Races of Man* by Carleton S. Coon, 1965, Table 9, p. 261. Copyright © 1965 by Carleton S. Coon. Reprinted by permission of Alfred A. Knopf, Inc.

TABLE 13.2 AVERAGE FREQUENCIES (EXPRESSED AS PERCENTAGE) OF
MAJOR BLOOD TYPES IN VARIOUS POPULATIONS*

Population Group	O	A	B	AB
Chinese	34.4	30.8	27.7	7.3
Japanese	30.1	38.4	21.9	9.7
Russians	31.9	34.4	24.9	8.8
Germans	36.5	42.5	14.5	6.5
French	39.8	42.3	11.8	6.1
Italians	45.9	33.4	17.3	3.4
English	47.9	42.4	8.3	1.4
Hawaiiens	36.5	60.8	2.2	0.5
Australian aborigines	42.6	57.4	0.0	0.0
American Indians (Utes)	97.4	2.6	0.0	0.0

*Adapted from *Human Genetics* by A. M. Winchester, 1971, Table 12-1, p. 165. Reprinted with permission of Charles E. Merrill Publishing Company.

vide clues to patterns of gene exchange between populations and of past migrations of human racial groups (Fig. 13.8).

Less reliable systems for distinguishing population groups include language, religion, cultural affiliation, social caste, economic group, or other congregations based on environmental rather than genetic factors and subject to change in numerous ways for numerous reasons. In fact, using established genetic factors one can show that allele frequencies vary within a cultural or ethnic group according to their racial components rather than their chosen identifications. Studies of blood groups or of frequencies of other genetic traits among Jewish populations reveal patterns associated with the geographical or racial origins of the groups and not with their religious affiliation (Table 13.4).

Migration of Human Populations

As mentioned earlier, past migrations of human populations would appear to provide the explanation for the spread of a species from its site of origin. It would be less likely that similar or identical groups arose independently in different parts of the world, especially in view of the time span between the localized and widespread distribution patterns. It seems likely that mankind originated in Africa, where the richest sources of australopithecines have been found and where *Homo habilis* lived, and spread from there to Asia and thence to Europe. Populations of *Homo erectus* are known from all three continents. Evolution toward *Homo sapiens* continued, perhaps via a generalized Neanderthaloid type, which progressed toward the sapiens form as well as producing a divergent "extreme" Neanderthal population.

TABLE 13.3   RANGE (IN PERCENT) OF BLOOD GROUP FACTORS
IN VARIOUS POPULATIONS*

Population	A₁†	A₂†	B	O	Rh⁺	Duffy factor
Caucasians	5–40	1–37	4–18	45–75	54–75	37–82
Negroes	8–30	1–8	10–20	52–70	71–96	0–6
East Asians	0–45	0–5	16–25	39–68	95–100	90–100
American Indians	0–20	0	0–4	68–100	100 (approx)	22–99

*Adapted from *The Living Races of Man* by Carleton S. Coon, 1965, Table 12, p. 286. Copyright © 1965 by Carleton S. Coon. Reprinted by permission of Alfred A. Knopf, Inc.
†The A₁ and A₂ blood antigens are distinguishable from one another, and each is a product of a different allele of the A-B-O gene.

It is not known whether all the kinds of humans are descended from one ancestral lineage or whether divergence occurred (perhaps in *H. erectus,* according to some) and each then evolved into the present races of the human species. The decision between monophyletic (one ancestral lineage for all races) or polyphyletic (multiple lineages, each diverging toward a major race) origins of the races cannot be made entirely on the basis of fossil evidence. We have few fossil specimens to reconstruct detailed histories from *Homo erectus* to *H. sapiens* over the hundreds of thousands of years that intervened. The human races comprise a single species fully capable of interbreeding and producing hardy offspring. If these genetically compatible groups represent an ancient divergence, it is difficult to explain

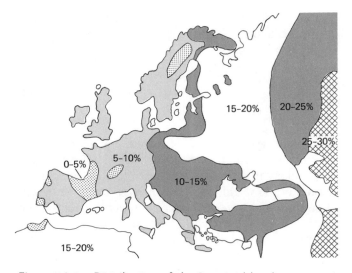

Figure 13.8   Distribution of the B major blood group antigen showing the apparent spread westward from an origin in central Asia. (Adapted from *Human Genetics* by A. M. Winchester, 1971, p. 165. Reprinted with permission of Charles E. Merrill Publishing Company.)

TABLE 13.4   BLOOD GROUP FREQUENCIES (IN PERCENTAGES) IN JEWISH AND NON-JEWISH POPULATIONS OF VARIOUS GEOGRAPHIC REGIONS*

Population from	Jews		Non-Jews	
	Type A	Type B	Type A	Type B
Africa				
Morocco	22.7	16.4	19.6	12.9
Algeria	21.5	15.5	20.6	12.6
Libya	22.7	16.4	20.4	11.6
Asia				
Yemen	17.4	9.5	17.5	6.7
Iran	25.9	18.5	22.7	16.0
Turkestan	20.8	21.6	26.5	18.5
Cochin (India)	11.6	14.3	19.0	14.4
Europe				
Netherlands	25.9	6.2	26.6	6.1
Germany	27.1	11.5	28.2	8.9
Poland	28.5	12.5	27.3	14.8
Ukraine	28.7	12.4	26.5	15.6
Lithuania	25.5	13.0	24.0	13.2

*Adapted from experiments of A. E. Mourant as tabulated by Th. Dobzhansky in *Mankind Evolving*, 1962, p. 241 (Yale University Press.)

how differences that arose in presapiens evolution later were reconstituted to form a common gene pool. If the racial divergence was initiated after *H. sapiens* appeared (which is the most acceptable genetic alternative), then the variations that distinguish the races are relatively recent. From our understanding of divergence in isolation it would seem logical to assume that present-day major races have accumulated some differences in allele frequencies, but that these variations are not so great as to create very different gene pools. We have evidence of the general anatomy and cultures of ancient populations, but such modern identifying traits as color, hairiness, facial characteristics, and similar racial guidelines are not preserved in the fossil record.

According to the archeological and paleontological evidence, a succession of cultures occurred during the last glacial and interglacial episodes, culminating in the Neolithic traditions of Africa and Asia. Prior to that time migrations had taken place from Asia to North and South America. The Indians of the Americas are descendants of an older Mongoloid group, showing various differences from the modern "new" Mongoloids, including the epicanthial fold of the eye. From the evidence it would seem that migrations to South America followed the arrivals into North America. Recent excavations in the Andes Mountains regions of South America provide indications of human cultures that may be 25,000 years old. If this is true, then the North American Indian ancestral cultures would be older than this.

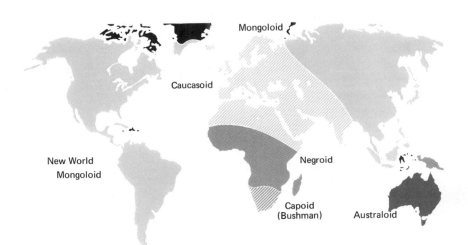

Figure 13.9   The distribution of the human species at the beginning of recorded history.

Human populations occupied all the continents and regions of the world, except for Antarctica, by the beginnings of recorded history (Fig. 13.9). All the races that are recognized today were in existence at that time. The African Negroids are a relatively young race, which probably arose after Capoids already were in existence, the latter being considered an ancient remnant group. Many new Mendelian populations appeared as migrations continued, and hybridizations undoubtedly occurred between indigenous inhabitants and entering groups. The intermingling of peoples today continues to produce new frequencies of alleles in the gene pool, as is especially evident in most recent times in Hawaii, Israel, and the continental United States.

Some Suggested Readings

Barnett, S. A., *The Human Species: A Biology of Man*. New York, Harper & Row, 1971, pp. 93–167.

Bleibtrau, H. K. (ed.), *Evolutionary Anthropology: A Reader in Human Biology*. Boston, Allyn & Bacon, 1969.

Garn, S. M., *Human Races*. Springfield, Illinois, C. C. Thomas, 1965.

Goldsby, R. A., *Race and Races*. New York, Macmillan, 1971.

Howells, W. W., "The distribution of man." *Scientific American* 203 (September, 1960), 112.

Howells, W. W., *Homo erectus. Scientific American* 215 (November, 1966), 46.

Krogman, W. M., "The scars of human evolution." *Scientific American* 185 (December, 1951), 54.

Laughlin, W. S., and R. H. Osborne (eds.), *Human Variation and Origins: Readings from the Scientific American*. San Francisco, Freeman, 1967.

MacNeish, R. S., "Early man in the Andes." *Scientific American* 224 (April, 1971), 36.

Mulvaney, D. J., "The prehistory of the Australian aborigine." *Scientific American* 214 (March, 1966), 84.

Napier, J., "The evolution of the hand." *Scientific American* 207 (December, 1962), 56.

Napier, J., "The antiquity of human walking." *Scientific American* 216 (April, 1967), 56.

Osborne, R. H., *The Biological and Social Meaning of Race*. San Francisco, Freeman, 1971.

Reed, T. E., "Caucasian genes in American Negroes." *Science* 165 (1969), 762.

Robinson, J. T., "Evidence for human evolution," in *Topics in the Study of Life: The BIO Source Book*. New York, Harper & Row, 1971, pp. 449–456.

Shapiro, H. L., "The strange, unfinished saga of Peking man." *Natural History* 80 (November, 1971), 8.

Simpson, G. G., "The biological nature of man." *Science* 152 (1966), 472.

Tobias, P. V., "Early man in East Africa." *Science* 149 (1965), 22.

Woolf, C. M., and F. C. Dukepoo, "Hopi Indians, inbreeding, and albinism." *Science* 164 (1969), 30.

Young, L. B. (ed.), *Evolution of Man*. New York, Oxford University Press, 1970, pp. 177–249.

CHAPTER 14

HUMAN
SOCIAL
ORGANIZATION

A feature common to the known anthropoid primates is the existence of social organization, which binds together the individuals and their activities in populations. In essence a society is considered to be a group that engages in activities experienced in common. Therefore, humans clearly are societal creatures along with their primate relatives. In bursts of literary analogy some authors have compared a society to a superorganism in the sense that there is a division of labor or of function or responsibility among the members of such a group. The analogy is not appropriate for all examples of social groups but may apply well enough to some sorts of animals. The comparison is made because a multicellular organism is a single individual comprised of hierarchies of cells, tissues, and organs in organizational levels. The whole population of cells functions cooperatively so that the individual accomplishes its many activities in an integrated and coherent fashion. But each component of the single organism loses its own individuality while it provides some essential contribution to the intact system. The organism is the individual and its component parts generally are incapable of independent existence and identity apart from the whole animal. The social insect societies, however, resemble the multicellular organism more closely than do the societies of more complex animal life. Among such insects as the ants, bees, wasps, termites, and others of these groups, each individual is a part of a rigidly defined system and each makes a particular contribution to that system. There is no individuality for any of the members of such an insect society because neither the individual nor the society exists outside the single framework of reference. In this case the animal society is analogous to a multicellular organism.

Social groups of more advanced species are comprised of individuals, and *individuality* has increased consistently during the evolutionary progression of animal groups (Fig. 14.1). In the case of humans and other social animals of more advanced groups, the behavior of the individual and the behavior of the group are regulated by a sort of feedback control. The individuality of the individual is accommodated to its existence in the social group by social constraints. The most important of these constraints are *status behavior* and *territoriality*. The status system permits individuals to stay together with little or no conflict. Dominance hierarchies (peck orders)

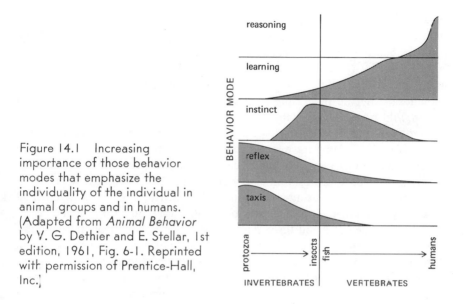

Figure 14.1 Increasing importance of those behavior modes that emphasize the individuality of the individual in animal groups and in humans. (Adapted from *Animal Behavior* by V. G. Dethier and E. Stellar, 1st edition, 1961, Fig. 6-1. Reprinted with permission of Prentice-Hall, Inc.)

are systems of social rankings, and these are maintained by individuals or by the group that monitors the individuals. In this way social life is not a free-for-all scramble for group resources; individual assertiveness is channeled and much reduced even though vigorous competitions may ensue for rank and its privileges. Territoriality refers to the subdivision of geographic space of the social group into domains, each of which is occupied and defended or maintained by individual members of the group. Individuals or whole groups thus are kept apart, with the consequence that aggressive disputes and competition for resources and mates are reduced. Dispersal over a large area is ensured by group territoriality. The human society is considered to comprise a mixture of hierarchical and territorial constraint systems, rather than depending exclusively on one or the other kind of constraint as may be the case in some other species. Hierarchies exist within the territories among the primates. The species therefore must contend with both forms of aggressive behavior, namely, (1) the acquisition of dominance in social hierarchies, and (2) the establishment of rights over particular pieces of geographical space or domains.

Studies of Various Animal Societies

Since the essence of a society consists in the behavior of its individual members in the maintenance of the social group, most of the literature has been concerned with particular aspects of these qualities. Studies have been concerned specifically with (1) societal roles assumed by individuals, (2) the behavioral traits shown by the group toward its members, (3) the bearing of group behavior on the maintenance of social group function, and (4) the development or evolution of societies in changing situations. Many studies of animal societies have utilized caged or zoo populations, but an increasing number of investigations have been carried out on wild populations in their natural surroundings.

Among nonprimate mammals the wolf has been studied extensively because its social groups have been considered to be quite similar in some respects to those presumed to typify early human or prehuman populations. In one particular series of studies conducted by J. H. Woolpy, a group of wolves was observed during a ten-year period in which it stayed together at Chicago's Brookfield Zoo. The group initially was comprised of five yearling cubs, two females and three males, which were siblings or half-siblings. The social group of these timber wolves (*Canus lupus*) consisted of an alpha-male, an alpha-female, subordinate males and females, and juveniles. In this dominance hierarchy the alpha-male was dominant over all members of the group and the alpha-female was dominant over all females and most of the males. The term "alpha" obviously refers to the status of greatest dominance for an individual. The peripheral males and females remained at some distance from the nucleus of the pack but tried to participate in pack activities as much as possible. Courtship and mating-pair formation largely were controlled by the social system of the pack. Males courted females and females also courted males. As has been observed also in wild packs of timber wolves, population size remained more or less constant owing to a form of "birth control." The patterns of courtship of the alpha-male and alpha-female, as well as their aggressive behavior, led to restrictions on matings of other pack members. The wolf does not mate for life, but the male may mate with the same individual for several years. If the alpha-male prevents others from mating, few or no other pairs may form within the social group, thus regulating the numbers of young introduced into the pack each year. This behavior pattern leads to lower diversity in pockets of the species gene pool because inbreeding occurs more often than random matings. But there is considerable diversity from one part of the species gene pool to another so that stored variability is present. Occasional matings between members of different groups can provide the opportunity for gene flow, which maximizes the genetic flexibility of such compartmentalized population structures.

Woolpy assumed that certain behavioral features were heritable and that others were nongenetic in expression. Since the social arrangement of the group was established at first by "untutored" yearlings and maintained afterward along expected lines for this species, Woolpy considered the social pattern to be genetically based. Mating behavior similarly was considered to be heritable, in view of the fact that the observed patterns were like those seen in all other social groups of this species. Individual social roles were not heritable, since there were exchanges in social roles within the group during the ten-year period. These interpretations are somewhat simplified. Potentials for development are inherited, and these potentials will be expressed in an appropriate environment, which elicits such responses of the genetic system. Some traits, such as blood type, will be expressed in any environment, but many complex traits are programmed for differentiation in relation to environmental influences during development. Behavioral traits generally are genetically programmed for flexibility of expression, at least in more advanced species groups. It seems that very few absolutes can be stipulated for higher animal behavior. Social relationships in wolf societies are formed early in life and nurtured by constant experience throughout the early years. This set of experiences permits, and is essential for, development of proper social responses and the integration of the young into adult society. In timber wolf populations the social groups apparently develop along predictable lines, whether or not socially experienced adults are present initially. The same situation probably prevails for primate social groups, but with greater degrees of flexibility and variability of pattern development.

The mountain gorilla has been studied in the wild by George Schaller, as described in his book covering a period of about one year during which he made short, day-long trips and occasionally stayed overnight near camps of gorilla groups. Schaller recorded from these observations that the gorilla social group was permissive and relaxed. There is a dominant male, usually older than others of similar size and maturity. The female invites the male to copulate when she is physiologically receptive, which is about three or four days of the month and coincident with ovulation (the estrus period). The male makes no overtures unless the female indicates receptivity. Because these gorilla groups are rather small (an average of 20, about half of which are adults) and because pregnant or lactating females are not sexually receptive, gorilla males may experience no sexual activity for as long as a year. The group is stable nonetheless. Sexual activity in this species appears to be of lesser importance to group stability than may be the case in other animals. The dominant male shares the females with other males of the group, or even with temporary visitors to the group. It was not known from this relatively short-term study whether such male visitors were strangers to the group or previous members, nor could it be determined whether such visitors could come and go with ease over long periods of time. In the presence of some danger the males alerted the group, using various sounds. All members of the group then scattered to the trees, including the dominant male; no members stayed behind actively protecting the group. Since all the members of the gorilla group forage for their own food among the plentiful vegetation, there appears to be no division of labor involved in providing for the food needs of the aggregate membership.

The chimpanzee seems to be a more active species than the gorilla, sexually as well as in other ways. Some studies of caged animals brought together for brief experimental periods have been described in strongly anthropomorphic and chauvinistic terms. These studies showed that the male exhibited dominant social behavior and also dominated females during courtship or mating. The male decided whether or not to copulate with the female, and he might refuse an opportunity even when the female presented herself to him. Presentations by the female appeared to have other functions in addition to invitation to copulate. Since copulation occurs more often among chimpanzees than among gorillas, sexual presentation may be practiced as a behavioral signal by the female to achieve some purpose unrelated to sex per se. The imputations of "wile" and "cunning" and "sexy" female behavior are more reflective of the point of view of the experimenter than related to the behavioral pattern of the chimpanzee. The human observer often implies the existence of some human motivational basis for behavior, for chimpanzees or other species, thus distorting the situation and obscuring an accurate appraisal of the pattern expressed.

Jane van Lawick-Goodall's studies of chimpanzees in their natural settings in central Africa have revealed many differences in behavioral patterns when compared with the artificial and distorting situations for most caged animals, especially when brought together without regard to previous history or social group context. Chimpanzees that are free-living behave in such a way that conflicts over females rarely occur; males approach females who are in estrus, and only very young females are observed to present themselves initially to males. As in the gorilla groups, chimpanzee male dominance was observed to be relatively relaxed in natural settings and not overbearing or neurotic, as may have been demonstrated in poorly designed experimental situations.

Social Group Patterns in Higher Primates

Although all primates live in social groups, few species have been studied in the wild. In addition to studies of the gorilla and chimpanzee already mentioned, there have been substantial reports of social behavior in natural populations of such Old World monkeys as langurs, macaques, and baboons. The size of the social group varies even between different species of the same genus; for example, chacma baboons live in groups of about a hundred members, whereas hamadryas baboons of the dry savannahs live in small families, which may include one male, several females, and juveniles up to the age of 18 months. Among the higher primates the male is dominant over all group members, and he defends either directly by signaling imminent danger or all clear, or by preventing individuals from straying in open-space habitats. Groups of males may act in concert in defense postures, and the location of group members while on the move or engaging in activities in a temporary location indicates defense arrangements very clearly (Fig. 14.2).

The alpha-male gets everything he wants and he gets it first; others wait their turn. There is little conflict for mates in many species and no regular intragroup conflict over food in these vegetarian societies. Changes in social status occur relatively often and sometimes are very temporary episodes. A subordinate member may have something a dominant individual wants, and the dominant will beg for it. A subordinate may attain a higher status by associating with a dominant individual, which may be a male or a female since females also exhibit dominance to subordinate males and juveniles as well as to other females.

Except for humans and gibbons, primates do not show permanent matings between pairs. Although the gibbon is monogamous, in different human societies various numbers of males and females may comprise the mating nucleus, with monogamy as the most frequent pattern. Female subhuman primates generally accept males only during estrus, although not as strictly as has been observed for other mammals or less advanced vertebrate species. Dominant males of some species have first choice at the height of receptivity of the female, but this situation varies among primate species. The baboon shows strict first choice by the dominant males, whereas the gorilla and chimpanzee do not.

Parental care is the most constant aspect of behavior serving to keep individuals together, in the form of parental and filial relationships. Sexual behavior may be an infrequent event; therefore it does not serve as a unifying bond in all higher primate social groups. Among the tree-living

Figure 14.2 Spatial relationships among baboons in a troop on the move. Dominant males, adult females, and young occupy the center of the group space; others are arranged in concentric circular formation. (Adapted from *Primate Behavior: Field Studies of Monkeys and Apes* edited by Irven DeVore, 1965, Fig. 3-10, p. 70. Copyright © 1965 by Holt, Rinehart & Winston, Inc. Reprinted with permission of Holt, Rinehart and Winston, Inc.)

langurs the females care for infants and the males appear to be indifferent to the young. Among macaques and baboons, on the other hand, males pay attention to the young and are permitted to help, even though on occasion the mother prevents other females from touching her offspring. Japanese macaque males are very solicitous toward yearlings in the group. The period of juvenile dependence and attachment varies, being one or two years for monkeys such as baboons and macaques and about three or four years for the gorilla and chimpanzee, respectively. The social development of the young is the result of complex sequences of social experience superimposed on a genetic background of potentials.

The fascinating studies by Harry and Margaret Harlow on monkeys raised in various social situations provide some indication of the complexities involved (Fig. 14.3). Young monkeys raised in isolation apparently may suffer permanent psychological damage and may never achieve a normal relationship with others of their species even if opportunities are presented later. Different levels of social integration and balance develop in monkeys raised only with peers as compared with parents, or with both peers and parents. Similarly, behavior patterns of different sorts become established for young monkeys raised with adult females and those raised

EXPERIMENTAL CONDITION	PRESENT AGE	BEHAVIOR*			
		none	low	almost normal	normal
Raised in isolation					
Cage-raised for 2 years	4 years	● ○ ●	●		
Cage-raised for 6 months	14 months	○ ●	●	● ○ ●	
Cage-raised for 80 days	10 months			● ○ ●	
Raised with mother					
Normal mother; no play with peers	1 year	●	●		○
Motherless mother; play in playpen	14 months			○	● ●
Normal mother; play in playpen	2 years				● ○ ●
Raised with peers					
Four raised in one cage; play in playroom	1 year				● ○ ●
Surrogate-raised; play in playpen	2 years				● ○ ●
Surrogate-raised; play in playroom	21 months				● ○ ●

* ● = play, ○ = defense, ● = sex

Figure 14.3   Summary of results from some of the experiments reported by Harlow and Harlow showing the importance of interactions between individuals and of learning for the development of effective social relations and species behavior in monkeys.

with surrogate mothers having different textural and feeding qualities. The development and establishment of adult social behavior reflects numerous influences throughout the early formative years of the individual. The patterns are *learned,* whether for sexual or for social activities.

Which of the subhuman primate groups should be selected as models for various human institutions, or as indicators of evolutionary relationship and progression? There are almost as many systems of mating, family, and dominance among primates as there are species. For a model of domestic tranquility one might choose the easygoing gorilla, the more active chimpanzee, or the ill-tempered but monogamous gibbon. The mother model could be chosen from two opposites within the same monkey genus: the stern pig-tailed macaque or the permissive stump-tailed macaque. Which father figure—the bossy baboon or the gentle gorilla? There are varied models for parental responsibility toward the young: the Japanese macaque father, which helps look after the babies; the marmoset father and mother, who take turns carrying their offspring; or the male chimpanzee, which leaves the job of raising the young to the mother and assumes little responsibility himself for his progeny. Should we select models from species that diverged 40 million years ago, such as the monkeys and gibbons, or from species such as the gorilla and chimpanzee, which diverged from the common ancestral line more recently in evolution?

Clearly, model selection would be completely arbitrary and of little value if considered out of context with life style. Distinctive differences exist between closely related species that encounter different stresses and challenges in their daily activities, such as the baboons of plentiful habitats as compared with those of hostile habitats. A more complete consideration of the human social group and its origins requires a close examination of human evolution and circumstances, as well as a look at subhuman societies for possible pathways of development and comparisons.

Sex Roles
in Human Societies

This topic is argued emotionally more often than it is discussed objectively, and data may be selected or interpreted arbitrarily to support some hypothesis. It would seem to be more profitable to analyze the information in support of various hypotheses very carefully before we try to understand human social behavior and its evolutionary history and premises. At this particular moment in the arguments, objectivity and defensiveness become intermingled and it becomes difficult to extricate one point of view from the other.

A theme underlying the ideas expounded by Robin Fox, Lionel Tiger, Desmond Morris, Robert Ardrey, and others is that social behavioral patterns in humans reflect an ancient heritage of hunting, which led to adaptations still present in our species. These particular traits are proposed to characterize societies of modern humans, since we are considered to be the genetic legacy of hunting ancestors and to have dropped the hunting pattern only in recent times. Three lines of evidence have been described in discussions concerning the division of labor in early hunting social groups, and that division presumably led to the delineation of particular sex roles in prehuman evolution—roles that have been perpetuated to the present day as a function of genetic heritage and continuity. The lines of evidence are (1) the fossil record, (2) comparative studies of primate behavior, and (3) cross-cultural patterns detected in modern human societies, especially among so-called primitive peoples.

The image of the dominant and capable male as opposed to the subordinate and inconsequential female in modern societies stems directly from the proposal that prehumans and early humans were hunters. Men required speed, skill, cunning, strength, endurance, bravery, discernment, and organization for the hunt, which provided the food that maintained the social group. Selection pressures acting on natural diversity thus would have led to an increased dichotomy between the two sexes; the successful males would be those who were modified appropriately in anatomy, physiology, and intelligence, and these traits would predominate in the gene pool. Females would have benefited only indirectly from such a gene pool, or not at all according to some authors. Hormonal controls would have become fixed in successful hunting groups, thus providing developmental regulation of desirable male attributes and an enhancement of the behavioral differences between the sexes. The rapid rate of gain in intelligence in hominids during the Pleistocene thus would be traced to the adaptive advantages of brains for the hunting males in the small communities; females would reap some reward only because they had access to this same gene pool.

It was also necessary for the successful hunters, on whom the whole social fabric depended in those times, to be released from concern over females and family, who stayed at the campsite; otherwise their concentration on the hunt would be impaired, to the detriment of the species. Selective advantages thus would have accrued for various modifications in the female, such as loss of estrus, a feature unique to human females among the primates. If the female could be sexually receptive at any time and not merely during the few days of ovulation each month, then the males could concentrate on the hunt at any time and remain away from the camp for longer intervals. The male could be confident that females would be serviced at the male's convenience and would not be impelled by the urges of estrus to seek or accept other males if he happened to be off hunting at the propitious moment. Furthermore, the female would be more likely to remain within his sphere of influence, even when he was away from home, because he could fulfill his family obligations at any time that he was free of other responsibilities. It also would follow from this line of argument that there would be no particular breeding season; breeding would be accomplished under new restraints controlled by the male and not by female physiology.

There would, however, be a new dilemma for the hunting male; although he would benefit in one way from the loss of estrus control over mating incidents, he might "worry" about the continuing receptivity of the female left at home while he was on a hunting expedition. Therefore, pair bonding of the male and female would have resulted, to further ensure that the female and the home would still be there and still belong to him when the male returned from the hunt. Pair bonding between particular males and females would also have helped to reduce frictions between competing males, thus leading to greater cooperativeness as well as concentration on the quarry among males engaged in big-game hunting.

As explained by Lionel Tiger, bonding between males also would be advantageous for hunters because this would provide further insurance of cooperativeness among males and lead to a steady food supply, which would benefit the entire social group. The management and decision-making activities would be male functions, since hunting provided the needs of the group, and the hunters were males upon whom the group depended. Skill and cooperation plus the expanding intelligence of the male hunters would have laid the genetic groundwork for the comparably significant monitoring of social needs and problems, at home as well as on the hunt. The social bonds required for the hunt, together with the intelligent manipulations of the home and community, thus would have provided the selective direction that fixed male genetic superiority into the fabric of the human social pattern during the evolutionary history of prehumans and humans.

In a hunting society females, too, would have undergone significant evolutionary changes, principally in relation to the needs of males as hunters. Intelligence was not required for bearing or rearing children; swiftness and strength were not required for food-gathering activities or for bustling around the campsite; and bonding between females was of little advantage because there would have been few group activities at home other than the gathering of food in nearby areas. Docility and the acceptance of a subordinate social role would have been favored, according to the hunter theorists, because the home had to be tranquil and stable for the males who bore the responsibilities for the community's existence. In summary, devotees of the male superiority syndrome and explainers of the naturalness of female subordinate status invoke the single theme based on the needs and advantages relative to the hunting males.

The fossil record clearly indicates that hominids engaged in hunting during the Pleistocene; deposits from Olduvai Gorge and elsewhere contain animal remains along with tools made of horn, bone, and stone. But these collections of australopithecine and habiline deposits provide no evidence for big-game hunting; rather the evidence indicates that these species had a hunting-and-gathering economy very similar to known practices of simple societies in modern times. Hunting activities more probably involved small and immature game, and such foods may not have comprised any more of the diet of the community than in modern times, when most of the food is secured by gathering rather than by hunting in "primitive" societies such as those of the Bushmen of southwest Africa. *Homo erectus*, on the other hand, was a big-game hunter as were the hominines that followed during an evolutionary sequence of 500,000 to 700,000 years, as attested by the fossil record. That record does not, however, provide information about the social significance of hunting; nor does it permit us to interpret the nature of the hunting party in relation to its size, the sex of its participants, or the duration of some expedition. One would assume that since females would be concerned with managing the infants and juveniles, males engaged in the hunt; but we only surmise what other features prevailed.

The hypothesis that a hunting primate existed in late Miocene to early Pliocene times is somewhat speculative. The fossil record provides some indication of a shift from a tropical forest to an open woodland habitat for *Ramapithecus* and thence to life on the open savannah for the first hominids. Along with this shift in habitat, upright bipedalism was initiated as a hominid trait and reduction occurred in the size of teeth. Therefore, the speculation is made that an alternative set of defenses would have been required to survive in more hostile terrain, and a new kind of diet would be required away from the tropical forest vegetation. The introduction of a hunting capacity is credited with solving both the problem of defense and the problem of diet. But other primates have adapted to savannah life without becoming predatory hunters, including the modern baboons and the extinct vegetarian pongids of the *Gigantopithecus* group. Of course, these forms did not develop upright bipedalism, or reduced canine teeth either, so there are differences to consider.

Many difficulties are encountered in trying to explain the sequence of events leading to the hominines. The same sorts of difficulties faced by those who subscribe to the hunter concept were considered recently by Elaine Morgan, who adopted another extremist view, quite different from the predatory prehuman image. She marshaled many levels of observation to provide a basis for proposing that prehominids pursued a semiaquatic existence during 10 million years of Pliocene drought, and that many of the traits that appeared rather suddenly in the Pleistocene species were a consequence of adaptations to the littoral life led in the preceding epoch. There is no more evidence for her ideas than for the hunting hypothesis, and, indeed, she commits the same mistakes by constructing a hypothesis and selecting data to fit the preconceived notions.

The dearth of Pliocene fossil evidence makes it almost impossible to substantiate any of the ideas that have been presented so far. With regard to sex roles, not one shred of evidence exists concerning social behavior or community patterns or the division of labor that characterized our ancestors. Fanciful pictures depicting the "home life" of prehumans and ancient humans are no more than artists' conceptions. They do not constitute evidence of a life style, much less of sex roles in social groups of fossil forms.

The relatively few studies of wild populations of higher primates do not permit us to make comparisons of social structure and sex role distribution in the groups. Even here it seems that the particular species selected for comparison or as models often prove to be monkeys rather than hominoids, that is, macaques and baboons rather than chimpanzees or gorillas. The Old World monkeys have been on a divergent course of evolution for at least 40 million years. The pongids themselves are quite different today from Miocene or Pliocene forms, about which we know very little indeed. The simian shelf of modern pongids appeared in more recent species, perhaps indicating behavioral adaptations unrelated to features that were present in the common ancestor of Pongidae and Hominidae. Arbitrary comparisons must be viewed with caution because we know little, generally, about the genetic patterns that underlie social behavior and sex roles, either in fossil forms or in existing species. Perhaps the basis for the frequent choice of the baboon as a model of prehominid or hominid sex role assignment has more to do with the fact that the male baboon shows strict dominance socially and sexually than with its being a savannah resident.

Studies of present-day "primitive" peoples have provided the evidence of cross-cultural similarities in distinctive sex roles for males and females. The implication is that a common social pattern exists in all human groups; therefore there must be a genetic basis to such expression, which must be of ancient origin since the pattern is universal. Analysis of cultures by social anthropologists frequently leads to controversial interpretations, and there is no general acceptance of the appropriateness or the accuracy of the descriptions that have been reported. There is no evidence of a phylogenetic continuity of cultural patterns from hominids to hominines to humans, or that the pattern in "primitive" societies today has been handed down unchanged from ancient human groups. Such assumptions are difficult to accept in view of the speed with which some of these societies have changed as a consequence of the appearance or encroachment of more technologically oriented groups in these same territories. Recent cultural changes have obscured ancient patterns of social behavior, according to our observations during the past 50 years. There are numerous examples of profound social changes that have occurred when peoples previously considered to be at a Paleolithic stage were exposed to twentieth-century people, ideas, and inventions. Such changes have swept away older cultural traits in one or two generations, as has been documented so well by Margaret Mead, among others; similar alterations could have happened in various places at various times to various cultures. Because of the selection of particular social patterns as models—usually with dominant males and subordinate females, since these are most frequent—the variability in patterns has been overlooked by many anthropologists who seek to explain sex role distribution in genetic and evolutionary terms.

One of the distinguishing features of human behavior is its plasticity and the lack of an inherent set of responses. It is by processes of socialization and acculturation that human individuals gradually acquire the habits, skills, and beliefs that integrate them into society. An infant is born with a genotype that permits receptiveness to conditioning by other human beings with whom it is associated, such as the family, the peer group, or society. Few traits are based on genetic information alone; most are a function of the interacting processes of heredity and environmental factors. The plasticity of human cultural behavior indicates that most of these traits are learned in response to societal conditioning. Any infant can become a member of any society, learning the language and customs of that society during the exposures and experiences of development. Depending on the interactions of genotype and experiences, any individual may be as intelligent, docile, strong, swift, and cooperative as the next individual of a similar constitution. A female growing up in New Guinea could develop any one of three temperaments relative to males in the same societies, according to Margaret Mead's studies. If they are in the Tchambuli tribe, the females will be dominant and the males subordinate; if Arapesh, then both females and males assume generally unaggressive and gentle behavior patterns; and if in the Mundugumor society, then females and males will be equally aggressive. Sex roles in society would thus seem to be more a product of societal conditioning and expectation and acceptance than coded in our genes as such, or in our "biogrammar" as Messrs. Fox and Tiger have proposed in a recent book (Fig. 14.4).

When we look at physical traits usually associated with dominance behavior, we see a decreasing dichotomy between males and females during prehuman and human evolution. Among the pongids or most of the monkeys there is a substantial difference in body size between males and females, and it is consistent throughout the species. Human males in a particular society may be slightly taller or heavier than females in that same society, but there is no consistency for these differences in the species as a whole. Individuals vary within each society, too, so we really cannot consider body size differences as invariant distinguishing traits for males and females in *Homo sapiens*. There is no difference in dentition; human males and females have canine teeth that are indistinguishable, unlike the situation in most of our primate relatives (Fig. 14.5). There is ample evidence of equivalent intelligence, educability, and brain chemistry and anatomy for human males and females. Obviously a number of physiological and ana-

*"I don't even like going bowling with the boys,
but it's programmed into my genes."*

Figure 14.4   (Drawing by Donald Reilly; © 1972 The New Yorker Magazine, Inc.)

tomical differences distinguish males and females, influenced by the sex hormone repertories present. In humans the masculinization of the embryo begins in about the sixth week and is completed by the end of the third month under the influence of fetal male hormones when the embryo is of the XY chromosome constitution.

According to recorded history and to observed social structures in human groups, males dominate in government of some societies and females in others. Equally successful or unsuccessful, totalitarian or open, warlike or peaceful, enlightened or narrow societies have been governed by men and by women during recorded history. In every case for which we have reasonably accurate and reliable information, there appear to be no invariant and absolute factors underlying differential sex role behavior in humans, other than the conditioning of social experience of each individual as an individual and as a member of a group. Either we will continue to perpetuate myths of superiority and inferiority of races, ethnic groups, age groups, and sexes, or we will bring our intelligence and reason to bear on

Figure 14.5 Relative size of the canine teeth in the baboon, a species that shows sexual dimorphism.

female   male

these insidious aspersions and permit each human being to achieve the dignity and level of contribution that is essential for our society to reap the full benefits of its potentials.

Some Suggested Readings

Alland, A., *Evolution and Human Behavior*. Garden City, New York, Natural History Press, 1967.

Alland, A., *The Human Imperative*. New York, Columbia University Press, 1972.

Fox, J. R., "The evolution of human sexual behavior," in *What a Piece of Work is Man* (eds., J. D. Ray and G. E. Nelson). Boston, Little, Brown, 1971, pp. 273–285.

Freedman, D. G., "A biological view of man's social behavior," in *Social Behavior from Fish to Man* (by W. Etkin). Chicago, University of Chicago Press, 1964, pp. 152–188.

Johnson, C. E. (ed.), *Contemporary Readings in Behavior*. New York, McGraw-Hill, 1970.

Mead, M., *Male and Female*. New York, Morrow, 1949.

Morgan, E., *The Descent of Woman*. New York, Stein & Day, 1972.

Morris, D., *The Naked Ape*. New York, McGraw-Hill, 1967.

Pfeiffer, J. E., *The Emergence of Man*. New York, Harper & Row, 1969.

Sherfey, M. J., *The Nature and Evolution of Female Sexuality*. New York, Random House, 1972.

Tiger, L., *Men in Groups*. New York, Random House, 1969.

Tiger, L., "The possible origins of sexual discrimination." *Impact of Science on Society* 20 (1970), 29.

Weisz, P. (ed), *The Contemporary Scene*. New York, McGraw-Hill, 1970, pp. 164–277.

CHAPTER 15

TRENDS IN HUMAN EVOLUTION

A unique quality of humans is the ability to dominate their biotic community. Unlike other species, humans can modify their societies, surroundings, and way of life. In the process we also may modify the existence of many other forms of life, from prokaryote to primate. The ability of human communities to modify their biotic surroundings began with the invention of agriculture about 10,000 years ago. Cultivating crops and domesticating animals not only permitted some freedom from hunting and gathering food, but also led to changes in the surrounding biotic community. Certain kinds of crop plants would lead to depletion of soil nutrients and a lower quality of the soil unless systems of crop rotation were instituted. Animal husbandry would modify the grazing areas and lead to changes in the kinds of wild plants and animals that could inhabit the same pastures. The institution of agriculture also permitted the urbanization of human society. And in urban concentrations there would be modifications in social groups, producing complexities that do not occur in small communities, whether of nomads, food gatherers, hunters, or agriculturists.

The history of the human species during its urban phase has seen the introduction of increasingly complex and technologically improved approaches to existence. With increased food supplies, more, and larger, populations could be supported. Improved living conditions became possible with improved technology and with the development of synthetics and manufactured products (wool clothing instead of fur wrappings, nylon fabric instead of cottons, etc.). These innovations have freed humans from total dependence on the particular resources in the immediate vicinity and allowed better utilization of dwindling resources, directly or in manufacture. Such changes have also led to abuse of the biotic community, to stresses of life in different societies, and to the many problems that have been recognized in modern civilizations. Unlike the other species with which we cohabit the planet, humans alone are capable of occupying many habitats and of filling a variety of living spaces, using cultural as well as biological inventions. The ability to dominate the biotic community is therefore as much a process of cultural as of biological evolution. These two phenomena are interdependent; together they form the basis for the pattern of human evolution.

Special
Qualities

The syndrome of qualities underlying the success of human evolution is a composite of intelligence and foresight, the ability to make and use tools, and social organization. The basis for social organization has a long evolutionary history. The social cohesiveness that characterizes all anthropoid primate societies serves as a framework within which additional qualities can be incorporated, while the whole social pattern itself can be modified as a consequence of learning and experience. Social organization has many forms in the human species, and the plasticity of organization permits a variety of life styles. This variety makes it impossible to describe universal behavioral traits or signals that can be recognized by all humans. The human social group is adaptable to many influences and conditions of existence, unlike the more proscribed patterns seen in subhuman primate societies.

A greater level of intelligence is required to manage in a social group that is flexible in behavior and in which learning is an essential process for successful integration into the complexities of social group membership. Selective advantages for increased intelligence may have led to the increased brain size and complexity required for cooperation within the group, a more subtle and demanding behavior than simple and limited responses to a finite set of audible and visual signals.

Robert Bigelow has proposed an interesting suggestion to explain the paradoxical characteristics of cooperativeness and aggressive warfare, which are distinctively human learned traits. His idea is that cooperation would have been an adaptive trait for prehuman groups living on the savannahs and at the mercy of predators (lions and leopards) that were larger, faster, and more efficient hunters. Protection of the young, without which the species cannot continue, required high levels of cooperation within prehuman groups—as we may observe today in subhuman primates living in open areas (see Fig. 14.2). At the same time the different groups would have competed for the richest resources, such as water, food, and living accommodations. Warfare between prehuman groups would have occurred, and those that triumphed presumably would have been the more intelligent. The brain trebled in volume in the 2 to 4 million years of evolution from *Australopithecus* to *Homo erectus* to *Homo sapiens,* which was a relatively rapid rate as compared with brain increase in the prehominid stages. Bigelow suggests that strong selection pressures would have been generated for the increased intelligence required for successful cooperativeness *within* the group and for successful warfare over resources, which went on *between* prehuman groups.

These same intelligent creatures would also have developed systems for *inter*group cooperation, according to Bigelow, thus gaining increasing intervals of peace during which existence still would be assured but fewer lives would be lost. But the cooperativeness within the group would still prevail because of strong pressures for the benefits that would accrue in daily life, and aggressive warfare would still occur spasmodically in competition for the best resources. Progressively larger and larger groups of individuals could be organized into societies because underlying *intra*group cooperation was the stronger selective theme, and increasing intelligence would have directed such organizational activity. This view of the evolution of human society is premised on the fixation of adaptively increased intelligence in prehuman groups and on a balance between intragroup cooperation and intergroup warfare, which together would provide the scene for the action of natural selection on diversity in the gene pool. Human behavior thus developed a substantial level of potential self-control.

But warfare also requires cooperation among the members of the groups that oppose each other and thus results from the capacity for highly organized activities, which can be modified in response to rapidly changing local conditions and events. As intelligence continued to increase owing to brain changes in evolution, there would have been more cooperation between some of the opposing groups so that some armies became larger than others, and larger concentrations of humans could coexist in cities, countries, and empires. The breakdown of large-scale society cooperativeness may have been followed by invasion and decimation by warring barbarian groups, as has occurred repeatedly during human society. Ultimately, however, order would be restored and cooperation would be pursued once more to build again a social system in which increasing numbers of people could live and contribute to group success. With more experience and control over all elements of social interaction, we may hope that the evolution of self-control will bring the human species to the stage of peace among all groups—in other words, to the realization that the advantages of cooperation outweigh the trophies of war, so that the human species may come to settle its intergroup problems without armies and holocausts.

Increase in intelligence and the acquisition of foresight, which comes with improved memory retention and greater brain capacity and complexity, clearly require a larger brain with more neurons. The striking difference in brain size between australopithecine and habiline species of 1 to 2 million years ago and the *Homo erectus* populations of 500,000 years ago brought the size of the prehuman brain to within the range that characterizes modern human beings:

Hominid form	Cranial volume (cm^3)	
	range	mean
Australopithecus species	435–600	500
Homo habilis	—	680
Homo erectus	775–1225	1000
Homo sapiens	900–2000	1300

The further increase in average cranial capacity occurred between the time of the last *H. erectus* populations through the Neanderthals to the first appearance of modern *H. sapiens* about 50,000 years ago.

Modern humans exhibit a substantial range of variation in brain size, which appears to be unrelated to intelligence *within* the species; that is, a person with a cranial volume of 1000 cm^3 may be as intelligent as someone whose brain volume is twice that amount. But average brain size differences between species, as between *H. erectus* and *H. sapiens,* probably are significant and representative of enhanced intelligence. Still, since a brain with a volume of 1000 cm^3 denotes a particular level of intelligence, it might seem that the increase in mean size in *H. sapiens* was superfluous, if we assume that intellectual equivalence prevailed in *H. erectus* and *H. sapiens* individuals with identical brain volumes. This assumption is based on the known toolmaking skills of prehuman and human species of *Homo* in the past 500,000 years and on measurements of intelligence in modern humans of different brain volumes.

Phillip Tobias has suggested an interesting explanation for the additional increase in brain size that occurred during the most recent phases of hominine evolution. Modern children attain a brain size of about 750 cm^3 at the end of the first year of life, and this is considered to be the minimum at which symbolizing ability becomes possible, according to some authorities. From the fossil evidence we know that this minimum size was not attained in *H. erectus* children until some time between the fourth and eighth year of life, with six years as the average for the species. If we accept these data and assumptions, then the implication is that a modern child can begin to acquire cultural experiences related to learning and intelligence at a very early age, including the ability to symbolize linguistically. The *H. erectus* child, on the other hand, was intellectually less capable until considerably later in development and therefore had less time for enculturation. According to this theory—in contrast with Bigelow's suggestion that selective forces favored brain increase, which raised the level of cooperativeness within the social group—Tobias suggests that the trend toward larger and more complex brains at least during the past 200,000 years has been toward an earlier capacity for enculturation and cognitive development. Such changes would lead to longer learning periods and longer intervals to incorporate experiences and learning between childhood and the introduction into society of the adult individual, who would be more fully developed experientially and intellectually. The skills of indirectness and symbolization, which constitute thinking as the behavioral quality of intelligence, have been studied by many investigators. According to Jean Piaget and others, these skills are developed over a period of 11 to 12 years.

Newer evidence indicates that symbolization by speech is a recent event in human evolution. Modern *Homo sapiens* is unique in communicating symbolically by language. The modern infant possesses a vocal apparatus that permits only incoherent sounds, which are phonetically inefficient and utilize few vowels. Human infants undergo an enlargement of the naso-pharyngeal tract during development; it is this large nasopharyngeal tract, together with the larynx or voice box, that provides the anatomical basis for coherent speech (Fig. 15.1). The adult pongid has an inadequate pharynx, and apparently Neanderthals also possessed anatomically deficient speech structures. The anatomical changes, coupled with the development of speech centers in the left hemisphere of the cerebrum of the brain and with the general increase in intelligence, all together provided the foundations for cultural exploitation, including characteristically human symbolized vocalizations.

Chimpanzees have been taught to utter a few simple words, sometimes after months or years of painstaking tutoring. For a long time their apparent inability to learn to speak was puzzling because their intelligence level theoretically was adequate for such learning. Now that it is known that chimpanzees are anatomically unsuited to uttering coherent words in any continuing fashion, the puzzle has been solved to our partial satisfaction. Chimpanzees have been taught to communicate symbolically using sign language. Thus despite their inability to vocalize coherently, these primates can learn intelligent symbolization up to a point. The experiments still are in progress, so we do not know yet just how complex the communication potential may be, but the basis for such a trait is present. Apparently there is relatively little sign or vocal communication among chimpanzees in the wild. It is not known whether a chimpanzee would be able to speak if it possessed the appropriate nasopharyngeal and laryngeal structures. Transplant operations might provide the answer to this question.

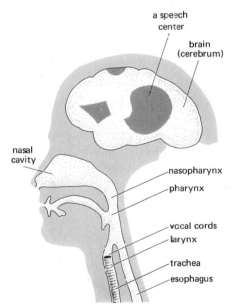

Figure 15.1   Some of the features of the brain and nasopharyngeal tract that contribute to the capacity for speech in humans.

Various concerns have been expressed about the lowering of intelligence in modern humans as a result of disproportionate production among different economic groups in societies. Intelligence is notoriously difficult to assess by the usual quantitative methods, but such IQ tests generally provide the evidence on which interpretations and conclusions are based. Most tests are not designed to test different generations of people, which would be necessary for any demonstration of a decline in intelligence with time. Where such a time factor has been introduced into testing, there has been a demonstrably significant increase in average test scores. There can hardly be a genetic basis to any change, in one generation or less, of a trait that is exceedingly complex in its informational and expressional parameters. These data more probably reflect changes in environmental factors, which contribute to test performances and results.

Mathematical calculations of the increase in deleterious alleles as a function of improved medical care and services, more relaxed social and legal regulations, and other factors all show that the rate of increase in such alleles is very slow indeed. Selection of advantageous alleles always will occur more rapidly and thus balance the increased genetic load of modern populations. So much of human behavior is a function of learning superimposed on hereditary potential that a general deterioration in health or intelligence would be a very unlikely consequence of the differences in reproduction. Even more to the point perhaps is the unproven contention that such populational changes are correlated with economic and social distinctions among people. To adhere to such a notion would be to imply that a genetic distinction leads to or underlies the social and economic class composition of societies. It is one thing to demonstrate a correlation between IQ level and economic or social group; it is quite another matter to impute a genetic basis to such a correlation. The difficulties of determining the relative input of heredity and environment on final phenotype expression of complex traits are all too evident in many of these studies. There are quantitative data showing clearly that human children mature earlier today than was the case two or three generations ago. We also have excellent data that demonstrate greater average longevity in many human populations over a span of a few generations. These modifications almost certainly can be traced to environmental factors, since genetic changes are not likely to occur so rapidly for multigenic traits in a species with a relatively long generation time, even when extremely strong selection pressures are operating.

Eugenics, Euphenics, and Genetic Engineering

Sir Francis Galton was a man of many interests and talents, and concepts he generated laid foundations that persist to the present day. One of Galton's interests was the possibility of improving mankind by selective breeding programs, for which he coined the term *eugenics*. The term has come into disrepute because of numerous occasions of abuse of the concept in its promulgation, and many of the distortions are clearly reprehensible. Racism, reactionary political views, and the heinous program of genocide practiced in Hitler's Germany are just a few of the examples of eugenics gone mad. Although the older and the distorted versions of eugenic measures are unacceptable, the new eugenics holds considerable promise, in theory.

Two programs that have received a mixed reception because of the complexities of management and differences in opinion and levels of confidence in the management are the suggestions for sperm storage banks and the more recently discovered possibilities of cloning cells.

Artificial insemination is in practice for humans, and the technology is under control. The eugenic program consists of storing sperm in deep-freeze banks and utilizing these gametes to produce improved human beings. Together with sperm obtained from donors considered to have desirable genes, there would be additional possibilities for fertilizing eggs of genetically desirable women outside the body and then transplanting the fertilized ovum into the uterus of a suitable proxy mother. The technology required for obtaining and storing ova, and for successful transplantation of the fertilized eggs to the uterus has not been completely worked out for humans. Such technical accomplishment may never be achieved because of the moral, social, and personal value problems involved. Furthermore, we cannot test individuals for the values that perhaps could be accepted unanimously by people today: values such as honesty, dedication, perseverance, integrity, and so forth. And we do not know what values might be considered desirable 10 or 50 or 100 generations from now, so in its long-term aspects, selective breeding tends to break down. Virtues tend to be viewed differently depending on which side one is on; for example, a soldier may receive the highest recognition for his bravery and success in war from his own country, but such awards are made for killing other human beings under conditions when murder is condoned by one or more of the warring factions.

Cloning presents very similar difficulties, technically and morally. It is possible to produce replicate individuals from single cells taken from an organism and grown in culture under controlled and optimum conditions. It is theoretically possible to obtain copies of individuals who carry "desirable" traits, and to maintain such replicates long after the original donor has died. Also, using the method of nuclear transplantation it has been possible to produce normal adult frogs and toads from eggs that have had their own nucleus inactivated or removed and a substitute diploid nucleus inserted instead. The genetic traits of the animals are encoded in the DNA of the transplanted nuclei, which can be obtained in the thousands or millions from any one individual's cells. In this way one can produce whole populations of exact replicas from eggs treated in this fashion. Whether such populations were created by cloning of cells or by producing replicas by nuclear transplantation, mass methods could be used to monitor populations with specific genetic instructions. It sounds like *Brave New World,* and like all the problems that would result from such a world.

A very promising eugenic program exists in the expanding areas of genetic counseling, which unfortunately is practiced all too little by the medical profession and for which there are limited service personnel and facilities. Advice can be given to prospective parents concerning the risk of having a child with a particular genetic defect, based on the hereditary features of the two individuals and their families. Thus if each of the prospective parents is found to be heterozygous for an inherited disease such as Tay-Sachs, they can be informed that there is a 25 percent risk of producing a child with the recessive genotype who would develop this lethal condition. The decision then is up to the two people as to whether or not to have a child. It also is possible in some kinds of genetic or chromosomal conditions to obtain a sample of the fetal cells by amniocentesis (Fig. 15.2) and determine directly whether the developing embryo is normal or not. For example, there is increasing risk of having a child with Down's syndrome (mongolism) for women who are over 40 years of age. If amniocentesis is performed early in pregnancy and 47 chromosomes are detected in the fetal cells, with chromosome-21 present as a trisomic (see Fig. 6.6), then the

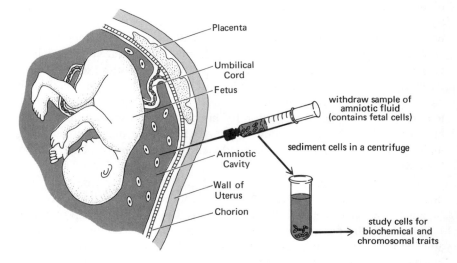

Figure I5.2   Amniocentesis. In the process of amniocentesis a sample of fluid is removed from the amniotic cavity in which the fetus floats, using a sterile hypodermic syringe. The sample is treated so that the fetal cells separate from the liquid portion of the amniotic fluid. A variety of tests then can be performed using the isolated fetal cells.

parents can be apprised of this situation and can decide whether to abort the pregnancy or be prepared for the birth of a child with mongolism. Here too there are problems of personal choice, legal complications, religious premises underlying decisions, and others that are equally grave. But the possibility of such medical services and advice exists for those who wish to avail themselves of them.

Euphenics refers to improvement of the phenotype using biological methods. The concept involves preventive and therapeutic medicine to correct genetic difficulties, which can be far more rapid and effective than eugenic measures. Children born with the recessive disease known as phenylketonuria lack an enzyme to process the amino acid phenylalanine, so that the amino acid accumulates along with another chemical that forms from some of the unmetabolized phenylalanine. These substances accumulate in the blood stream and lead to toxic effects on the central nervous system and to the symptoms of the disease. The disease can be detected in infants, whereupon the usual diet of milk and eggs and other foods high in phenylalanine is replaced by a special low-protein diet plus fruits and vegetables. Many problems are involved in the proper diagnosis, treatment, and evaluation of this disease, but it can serve as an example of the amelioration of a genetic defect, since such children do not develop severe mental retardation in many cases when the special diet is instituted in infancy. But there is no change in the gene or genotype, and the recessive alleles can be transmitted unchanged to progeny of such salvaged human beings. Treatment of afflicted individuals using hormones, enzymes, antigens, and other chemicals that cannot be manufactured or used by genetically altered individuals also can modify the phenotype and permit such people to lead useful, productive, and comfortable lives. Even organ transplants and the use of artificial organs come within the purview of euphenic measures.

Genetic engineering involves the direct modification of the encoded information and constitutes a new kind of possibility, which relies very heavily on advances in molecular biology and molecular genetic studies. It has been possible to transfer specific genes from one cell lineage to another by one of a number of methods, such as transfer via a virus vector or direct transformation under the influence of pure DNA incorporated into a host cell and then directing new phenotype expressions from the introduced coded information. It should be possible to introduce a functional gene capable of directing the synthesis of a missing enzyme, as an alternative to providing the enzyme directly or modifying the diet or living conditions of an afflicted individual. In such a case the functional gene that was added to the genotype could be transmitted in place of the defective gene for which it was substituted in the chromosome. Of course, such methodology is not yet within reach for human genetic problems but there have been successful experimental results with other kinds of cells and organisms. It remains a distinct possibility for future generations.

All these methods have been tested on lower organisms, and such research does not impose the same moral, social, ethical, and political overtones as would arise with genetic applications to human beings. There are prospects for harm as well as for good, for use and for abuse, and there is the lurking specter of a controlled society with all its sinister implications. But until we have explored and exploited any means for the alleviation of human misery by biological methods, we should not shy away altogether from a full evaluation of the potentials for improvement and for benefit that lie in the application of biological knowledge.

The fantastic potential for evolution, culturally as well as biologically, has led the human species to powers that have never been possible or realized before. We can swim without evolving fins, cross oceans and mountains in hours, fly without the aid of wings, inhabit regions of constant or of fluctuating temperatures, and perform many activities for which we have no biological attributes other than intelligence and toolmaking abilities. The possibilities for the future rest with us to a large degree and not with chance alone, as with other species. If we survive it will be due in large measure to our intelligence; if we extinguish ourselves it will be due primarily to our arrogance and stupidity in managing our affairs. We have, moreover, arrived at a state that carries with it the grave responsibility for the continuation of the planet itself, which we share with millions of other living forms that depend on one another and on the resources we all require for continued existence.

Some Suggested Readings

Bigelow, R., *The Dawn Warriors: Man's Evolution Toward Peace*. Boston, Atlantic–Little, Brown, 1969.

Bigelow, R., "Relevance of ethology of human aggressiveness." *International Social Sciences Journal* 23 (1971), 18.

Davis, B. D., "Prospects for genetic intervention in man." *Science* 170 (1970), 1279.

Friedmann, T., "Prenatal diagnosis of genetic disease." *Scientific American* 225 (November, 1971), 34.

Li, C. C., "Human genetic adaptation," in *Essays in Evolution and Genetics* (M. K. Hecht and W. C. Steere, eds.). New York, Appleton, 1970, pp. 545–577.

Premack, A. J. and D. Premack, "Teaching language to an ape." *Scientific American* 227 (October, 1972), 92.

Stebbins, G. L., "The natural history and evolutionary future of mankind." *American Naturalist* 104 (1970), 111.

Tanner, J. M., "Earlier maturation in man." *Scientific American* 218 (January, 1968), 21.

Tobias, P. V., *The Brain in Hominid Evolution*. New York, Columbia University Press, 1971.

APPENDIX A

The five-kingdom system outlined here is adapted from the proposal of R. H. Whittaker in his publication "New Concepts of Kingdoms of Organisms," published in *Science,* volume 163, 1969, pp. 150–160. There are other concepts of classification in addition to the one given here, such as the two-kingdom system, which places all organisms in either the Plant or the Animal Kingdom, and a four-kingdom system suggested by H. F. Copeland in his book *The Classification of Lower Organisms* (Pacific Books, Palo Alto, California, 1956) and two earlier publications. Copeland's system is similar in many ways to the one proposed by Whittaker, but one significant difference is that Whittaker separates the Fungi as a discrete Kingdom of organisms while Copeland includes the Fungi in the Protista (his precise term was Protoctista). Regardless of the particular system used there is internal consistency to the groupings and a logical set of criteria for classification. There is no agreement, however, on the one system that best suits the variety of living forms as endproducts of evolutionary pathways.

Kingdom
Monera

Organisms lacking nuclear membranes, mitochondria, chloroplasts, or complex flagella; prokaryotic cells not organized into tissues; varied nutritional mechanisms but excluding ingestion; reproduction usually by asexual means; some sexual recombination mechanisms known for a few species.

Principal groups of organisms are the bacteria and the blue-green algae.

Kingdom
Protista

Primarily unicellular or colonial-unicellular organisms with eukaryotic cells that contain nuclear membranes, mitochondria, and other membrane-bound organelles; diverse modes of nutrition including absorption, ingestion, and photosynthesis or some combination of these; true sexual processes in most groups, including nuclear fusion and meiosis; complex flagella or cilia, when present.

Many kinds of organisms, including euglenoids, golden algae and diatoms, dinoflagellates, and a variety of protozoa such as ciliates, sporozoans, rhizopods, suctorians, and animal flagellates.

Kingdom Fungi

Primarily multinucleate organisms with eukaryotic cells generally not organized into complex tissues except for reproductive structures in some groups; absorptive nutrition exclusively; both sexual and asexual processes included in reproductive cycles; plastids and photosynthetic pigments lacking.

Principal groups are the true fungi which have tubular cell organization (mycelium) and the slime molds. True fungi include water molds, conjugation fungi, sac fungi, and club fungi (mushrooms and others), along with the imperfect fungi, which lack known sexual phases.

Kingdom Plantae

Multicellular organisms with cells that have rigid walls, usually photosynthetic; primarily nonmotile, living anchored to a substrate; specialized tissue development related to conduction of water and foods, anchorage, support, reproduction, and covering tissues; primarily sexually reproducing with alternating haploid and diploid phases, the haploid phase becoming greatly reduced in the more advanced groups of plants.

Principal groups include three phyla of algae (red, brown, and green algae), and the primarily terrestrial bryophytes (mosses and liverworts), which lack vascular tissues (water- and food-conducting xylem and phloem, respectively); plus the tracheophyte terrestrial forms, which have vascular tissues. Tracheophyte groups include psilophytes, lycopods, horsetails, and ferns, all of which do not produce seeds; cycads and conifers of the gymnosperm groups, which produce naked seeds (not enclosed in maternal tissue of a fruit); and the angiosperms or flowering plants, which bear seeds enclosed in fruits. The number of phyla varies from one system of classification to another, and the specific phylum designations also vary (for example, flowering plants are called Angiospermae or Anthophyta, depending upon the system in use).

Kingdom
Animalia

Multicellular organisms with eukaryotic cells that lack walls, plastids, and photosynthetic pigments; nutrition primarily ingestive, but some forms are absorptive; exceedingly elaborate tissue and organ differentiation in many groups, more than in any other kingdom of organisms; predominantly sexual reproduction, with haploid phase generally restricted to the gametes.

Some systems of classification include Protozoa and Porifera (sponges) in this kingdom, but Whittaker's system considers them to be protists. Other phyla of animals include:

Phylum Coelenterata (corals, anemones, *Hydra*, and the so-called jellyfish)
Phylum Ctenophera (comb jellies or "sea walnuts")
Phylum Platyhelminthes (flatworms such as *Planaria*, flukes, tapeworm)
Phylum Nemertea (proboscis worms)
Phylum Nematoda (roundworms such as pinworm, hookworm, trichina worm)
Phylum Rotifera (smallest of the metazoans, including the rotifers or "wheel animals")
Phylum Bryozoa (moss animals of microscopic size)
Phylum Brachiopoda (lampshells and other marine animals with two hard shells that superficially resemble clams; most representatives now extinct)
Phylum Annelida (the segmented worms such as earthworms, sandworms, leeches)
Phylum Onycophora (rare tropical animals structurally intermediate between the annelids and the arthropods)
Phylum Arthropoda (segmented animals with jointed appendages and a hard outer skeleton of chitin; body divided into head, thorax, and abdomen).
Classes of arthropods include:
Trilobita (marine forms that became extinct in the Permian period)
Crustacea (crayfish, lobster, shrimp, barnacle, and others)
Chilopoda (centipedes, having one pair of appendages per body segment)
Diplopoda (millipedes, having two pairs of appendages per body segment)

Arachnoidea (spiders, mites, scorpions, ticks, all with four pairs of legs)

Insecta (many forms, all having three pairs of legs)

Phylum Mollusca (the unsegmented, soft-bodied animals usually but not always covered by a shell, such as clams, oysters, mussels, snails, slugs, squid, octopus, and others)

Phylum Echinodermata (marine forms that include starfish, sea urchins, sea cucumbers, sea lilies, and other forms)

Phylum Chordata (see Chapter 12)

Subphylum Hemichordata (acorn worms)

Subphylum Urochordata (tunicates and sea squirts)

Subphylum Cephalochordata (*Amphioxus* and lancelets)

Subphylum Vertebrata (fish groups and the four-footed terrestrial classes)

INDEX